Energy Consumption at Production Stages in Mining

Energy Consumption at Production Stages in Mining

Guest Editors

Krzysztof Skrzypkowski
Krzysztof Zagórski

MDPI

Basel • Beijing • Wuhan • Barcelona • Belgrade • Novi Sad • Cluj • Manchester

Guest Editors
Krzysztof Skrzypkowski
Faculty of Civil Engineering
and Resource Management
AGH University of Krakow
Krakow
Poland

Krzysztof Zagórski
Faculty of Mechanical
Engineering and Robotics
AGH University of Krakow
Krakow
Poland

Editorial Office
MDPI AG
Grosspeteranlage 5
4052 Basel, Switzerland

This is a reprint of the Special Issue, published open access by the journal *Energies* (ISSN 1996-1073), freely accessible at: https://www.mdpi.com/journal/energies/special_issues/T4AFTM4441.

For citation purposes, cite each article independently as indicated on the article page online and as indicated below:

Lastname, A.A.; Lastname, B.B. Article Title. *Journal Name* **Year**, *Volume Number*, Page Range.

ISBN 978-3-7258-5221-5 (Hbk)
ISBN 978-3-7258-5222-2 (PDF)
https://doi.org/10.3390/books978-3-7258-5222-2

Contents

About the Editors

Krzysztof Skrzypkowski

Krzysztof Skrzypkowski is an Associate Professor of mining engineering at the AGH University of Krakow, Faculty of Civil Engineering and Resource Management, Department of Mining Engineering and Occupational Safety in Poland. Since 2021, he has been working as a University Professor. His scientific research concerns the stability of underground mine workings, particularly the selection of support for room and roadway excavations. In addition, his research interests focus on energy consumption in underground mining, computer-aided design and determining the geotechnical parameters of backfilling materials. As the head of the rock bolting laboratory, he performs model, numerical, and industrial tests on various cooperation mechanisms between mining supports and rock masses. He is the author of over 200 scientific publications, several patents, and a utility model regarding artificial pillars, especially interrooms, support housing, a load absorber of the rock bolt support, a support pillar for a mine roof, a clamp for a mining anchor, an anchor plate system with a bolt load indicator, yielding a rock bolt support load indicator, a clamp for a mining anchor, a crib in the mining industry, a method for measuring the anchor loading and the dynamometric anchor washer, and a liable supporting foot of the mining longwall lining. As an academic lecturer, he works closely with scientific institutes, industries, and universities involved in exploiting mineral and rock raw materials.

Krzysztof Zagórski

Krzysztof Zagórski is an Assistant Professor of mechanical engineering at the AGH University of Krakow, Faculty of Mechanical Engineering and Robotics, Department of Manufacturing Systems in Poland. His research concerns the assessment of the wear of industrial tools and devices, and the design and manufacturing of machine elements. His research interests are also focused on destructive and non-destructive methods of structural elements. As a design engineer, he participates in the design of innovative machines and devices that are characterized by a reduced level of energy consumption while meeting strength, functionality, and ergonomic requirements. He is the author of over 130 scientific publications and several patents, concerning variable-speed transmission for machines with controlled motor power, a tool for thread forming, a clamp for a mining anchor, a drilling tool, a method for measuring the anchor loading and the dynamometric anchor washer, and a spindle with automatic tool exchange. As an academic lecturer, he closely collaborates with research institutes, industry members and universities dealing with engineering manufacturing systems.

energies

MDPI

Article

Load and Energy Consumption Analysis of Rope Winches Used in Mining Rail Transport

Krzysztof Krauze [1], Ryszard Klempka [2], Kamil Mucha [1,*] and Tomasz Wydro [1]

1 Faculty of Mechanical Engineering and Robotics, AGH University of Krakow, al. Mickiewicza 30, 30-059 Kraków, Poland; krauze@agh.edu.pl (K.K.); wydro@agh.edu.pl (T.W.)
2 Faculty of Electrical Engineering, Automatics, Computer Science and Biomedical Engineering, AGH University of Krakow, al. Mickiewicza 30, 30-059 Kraków, Poland; klempka@agh.edu.pl
* Correspondence: kmucha@agh.edu.pl

Abstract: A central engineering concern in the use of such winches is their energy intensity, which is significantly influenced by operational factors such as the length and gradient of the transport route, the transported weight, and the target hauling velocity. These variables directly affect the drive system's load and thus the power demand of the motor, which represents the primary component of the system's energy consumption. While manufacturers typically provide structural, kinematic, and power-related specifications, the user must evaluate whether a given winch can operate efficiently within the energy constraints imposed by specific site conditions. To aid in this assessment, an analytical model, along with a computational software tool, has been developed. This solution enables the continuous determination of essential parameters, with particular attention paid to the energy requirements of the system, supporting informed selection and implementation of winches with appropriately rated motors to optimize energy consumption. For the analyzed case, it was calculated that doubling the transport distance results in an increase in energy consumption of approximately 50%. Moreover, an increase in the velocity of the transported weight by only 0.5 m/s leads to a rise in energy consumption of approximately 25%.

Keywords: underground mining; rail transport; rope winch; drive load; energy consumption

1. Introduction

For decades, the transport of various types of machinery, equipment, materials, and personnel has been conducted using traction drives based on ropes and winches [1]. This type of transport system is particularly widespread in underground horizontal and inclined tunnels (Figure 1) [2–4], as well as in vertical shafts of underground mines (Figure 2) [5,6].

This method involves the movement of loads using a rope or occasionally a chain as the traction element, where the transport vehicle is guided either by rail tracks or mechanical guides [7,8]. One typical example of such transport systems is rail haulage utilizing winches equipped with ropes. Most often, the transportation of materials or personnel in rail-wheel-based cars occurs along horizontal or inclined transport routes.

Such a transport system generally consists of a single- or double-drum winch, a rope, one or more transport cars with rail chassis, a return pulley with a rope tensioning mechanism, and the track system itself. In inclined underground tunnels at four degrees or more, additional safety devices are installed to prevent the uncontrolled rolling of transport vehicles in the event of rope or winch failure [9,10]. In contrast, in tunnels with an inclination of up to four degrees, such protective devices are not required and therefore not used. It is evident that the rope winch in this system serves as the primary drive unit [11,12].

For a prospective user considering the implementation of a rope winch transport system, key drive parameters are of interest, such as the load hauling speed, the maximum rope tension force, and the motor power. These parameters must be determined either analytically or empirically as functions of the transport route characteristics and the weight of the material to be conveyed. Manufacturers of this type of equipment do not provide calculation methods related to winch-based transport. Therefore, the aim of this article is to assist users of such systems in the process of calculating and selecting their basic parameters.

Figure 1. Transport system with a rope winch in an inclined underground tunnel.

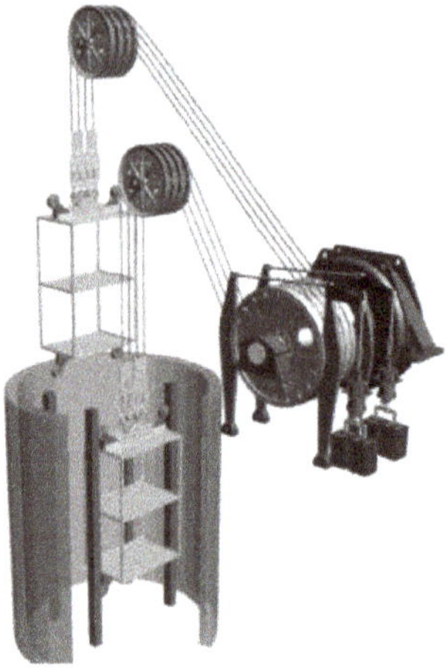

Figure 2. Hoist model with the rope winch [13].

2. Analytical Model of a Transport System with a Rope Winch

As outlined earlier, the proper selection of a rope winch requires the precise specification of its technical characteristics, which is most effectively achieved through analytical evaluation. This creates the need to formulate a dedicated analytical model for the transport system incorporating a rope winch. A conceptual representation of such a system operating along an inclined mining route is shown in Figure 3. In the proposed configuration, the transport process is driven by a winch operating with a continuous (endless) rope. The rope may be wrapped around the winch drum multiple times in order to generate the necessary frictional contact (XP) between the rope and the drum. Alternatively, if the system design assumes that the rope or ropes are mechanically fixed to the drum, the frictional interaction can be disregarded in the analysis.

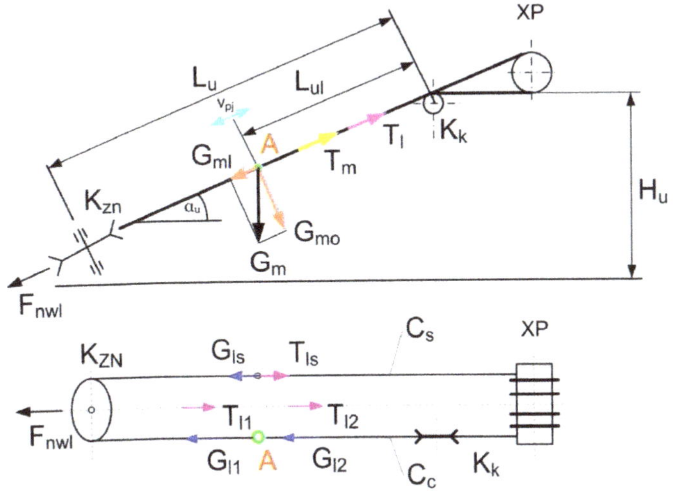

Figure 3. Schematic diagram of rail transport in an inclined underground tunnel using a rope winch: K_{ZN}—return and tensioning drum, K_K—guiding drum, C_C—winding rope, C_S—unwinding rope, H_u—vertical distance between K_{ZN} and K_K, v_{pj}—transport velocity, L_u—transport distance, L_{ul}—distance from arbitrarily selected point A to the guiding drum K_K, α_u—tunnel inclination angle, F_{nwl}—initial rope tension force, G_m—transported weight, G_{ml}—component of transported weight parallel to the floor, G_{mo}—component of transported weight perpendicular to the floor, G_{l1}—component of the rope weight section from K_{ZN} to point A parallel to the floor, G_{l2}—component of the rope weight section from point A to the drive unit parallel to the floor, G_{ls}—component of the unwinding rope weight parallel to the floor, T_m—resistance force for the movement of transported weight, T_l—resistance force for the movement of the winding rope, T_{l1}—resistance force for the movement of the rope section from K_{ZN} to point A, T_{l2}—resistance force for the movement of the rope section from point A to the drive unit, and T_{ls}—resistance force for the movement of the unwinding rope.

Then, using the principle of equilibrium on an inclined plane, the individual forces relevant to selecting the winch, rope, and rope tensioning mechanism were determined.

$$G_{ml} = G_m \cdot \sin \alpha_u \quad (1)$$

$$G_l = G_{l1} + G_{l2} \quad (2)$$

$$G_{l1} = q_l \cdot (L_u - L_{ul}) \cdot \sin \alpha_u \quad (3)$$

$$G_{l2} = q_l \cdot L_{ul} \cdot \sin \alpha_u \quad (4)$$

$$G_{mo} = G_m \cdot \cos \alpha_u \quad (5)$$

$$G_{ls} = q_l \cdot L_u \cdot \sin \alpha_u \quad (6)$$

$$T_{m} = G_{m} \cdot \mu_{pm} \cdot \cos \alpha_{u} \quad (7)$$

$$T_{l1} = q_{l} \cdot \mu_{pl1} \cdot (L_{u} - L_{ul}) \cdot \cos \alpha_{u} \quad (8)$$

$$T_{l2} = q_{l} \cdot \mu_{pl2} \cdot L_{ul} \cdot \cos \alpha_{u} \quad (9)$$

$$T_{l} = T_{l1} + T_{l2} = q_{l} \cdot \mu_{pl1} \cdot \cos \alpha_{u} \cdot (L_{u} - L_{ul}) + q_{l} \cdot \mu_{pl2} \cdot \cos \alpha_{u} \cdot L_{ul} \quad (10)$$

$$T_{ls} = q_{l} \cdot L_{u} \cdot \mu_{pl} \cdot \cos \alpha_{u} \quad (11)$$

$$F_{s} = G_{ls} \pm T_{ls} + 0.5 \cdot F_{nwl} = 0.5 \cdot F_{nwl} + q_{l} \cdot L_{u} \cdot \left(\sin \alpha_{u} \pm \mu_{pl} \cdot \cos \alpha_{u} \right) \quad (12)$$

$$F_{C} = 0.5 \cdot F_{nwl} + G_{lmC} \mp T_{m} \mp T_{l} \quad (13)$$

$$G_{lmC} = G_{ml} + G_{l} = (G_{m} + q_{l} \cdot L_{u}) \cdot \sin \alpha_{u} \quad (14)$$

where

q_l—unit weight of the rope [kN/m];
G_l—component of the winding rope weight parallel to the floor [kN];
μ_{pm}—rolling resistance coefficient for the transported weight;
μ_{pl}—rolling resistance coefficient for the unwinding rope;
μ_{pl1}—rolling resistance coefficient for the winding rope section from KZN to point A;
μ_{pl2}—rolling resistance coefficient for the winding rope section from point A to the drive unit;
F_C—tension force in the winding rope [kN];
F_S—tension force in the unwinding rope [kN].

From the above relationships, it is possible to determine the magnitudes of the relevant forces, with particular emphasis put on the tension in the winding and unwinding segments of the rope. This, in turn, enables the appropriate selection of the rope diameter d [14,15]. By specifying the transport velocity of the system and the weight of the material being conveyed, the required drive motor power N for the winch can be calculated [16,17]. The rolling resistance coefficients μ_{pl1}, μ_{pl2}, and μ_{pl} are typically introduced as empirical values, determined on the basis of practical assumptions. As a result, two scenarios may be considered. In the first, it is assumed that all coefficients take on equal values, in which case the following occurs:

$$\mu_{pl1} = \mu_{pl2} = \mu_{pl}$$

$$T_{l} = q_{l} \cdot \mu_{pl} \cdot L_{u} \cdot \cos \alpha_{u} \quad (15)$$

$$F_{C} = 0.5 \cdot F_{nwl} + (G_{m} + q_{l} \cdot L_{u}) \cdot \sin \alpha_{u} \mp G_{m} \cdot \mu_{pm} \cdot \cos \alpha_{u} \mp q_{l} \cdot \mu_{pl} \cdot L_{u} \cdot \cos \alpha_{u}$$

$$F_{C} = 0.5 \cdot F_{nwl} + G_{m} \cdot \sin \alpha_{u} + q_{l} \cdot L_{u} \cdot \sin \alpha_{u} \mp G_{m} \cdot \mu_{pm} \cdot \cos \alpha_{u} \mp q_{l} \cdot \mu_{pl} \cdot L_{u} \cdot \cos \alpha_{u}$$

$$F_{C} = 0.5 \cdot F_{nwl} + G_{m} \cdot (\sin \alpha_{u} \mp \mu_{pm} \cdot \cos \alpha_{u}) \mp q_{l} \cdot L_{u} \cdot \left(\sin \alpha_{u} \mp \mu_{pl} \cdot \cos \alpha_{u} \right) \quad (16)$$

In the second case, these coefficients have different values, and the lengths of the rope segments before and after point A also differ.

$$\mu_{pl1} \neq \mu_{pl2}, \; L_{u} \geq L_{ul}, \; L_{ul} \neq const$$

$$T_{l} = q_{l} \cdot \cos \alpha_{u} \cdot \left[\mu_{pl1} \cdot (L_{u} - L_{ul}) + \mu_{pl2} \cdot L_{ul} \right] \quad (17)$$

$$F_C = 0.5 \cdot F_{nwl} + (G_m + q_l \cdot L_u) \cdot \sin\alpha_u \mp G_m \cdot \mu_{pm} \cdot \cos\alpha_u \mp q_l \cdot \cos\alpha_u \cdot \left[\mu_{pl1} \cdot (L_u - L_{ul}) + \mu_{pl2} \cdot L_{ul}\right]$$

$$F_C = 0.5 \cdot F_{nwl} + G_m \cdot (\sin\alpha_u \mp \mu_{pm} \cdot \cos\alpha_u) \mp q_l \cdot L_u \cdot \sin\alpha_u \mp q_l \cdot \cos\alpha_u \cdot \left[\mu_{pl1} \cdot (L_u - L_{ul}) + \mu_{pl2} \cdot L_{ul}\right] \quad (18)$$

The use of a rope for transporting machinery, equipment, materials, and personnel in systems employing endless-rope winches [18] necessitates the implementation of a tensioning mechanism capable of applying the required initial rope tension force F_{nwl}. The value of this force is determined by the frictional coupling between the rope and the drum and is directly related to the tension in the pulling rope segment. The initial rope tension force F_{nwl} is typically either constant or adjustable, depending on the operational load. However, due to elongation or contraction of the rope under varying load conditions, compensation must be provided by the rope tensioning mechanism. Therefore, such a mechanism must be capable of delivering both an appropriate stroke length ΔF_{nwl} and a corresponding initial rope tension force F_{nwl}.

Assuming that ΔF_C represents the elongation of the rope under the action of the winding force, and ΔF_S the elongation under the action of the unwinding force, and that q_{lw} denotes the assumed unit elongation of the rope, the following relationship can be established:

$$\Delta F_C = q_{lw} \cdot F_C \quad (19)$$

$$\Delta F_S = q_{lw} \cdot F_S \quad (20)$$

Then, the elongation of the rope ΔF_{nwl} under the action of the initial rope tension force F_{nwl} is determined using Equation (22). In contrast, the elongation of the rope under the action of winding and unwinding forces can be obtained from Equation (21). The sum of these displacements corresponds to the minimum stroke of the return drum in the rope tensioning mechanism, as given by Equation (23).

$$\Delta F_{CS} = \Delta F_C + \Delta F_S = q_{lw} \cdot (F_C + F_S) \quad (21)$$

$$\Delta F_{nwl} = q_{lw} \cdot F_{nwl} \quad (22)$$

$$\Delta F = \Delta F_{CS} + \Delta F_{nwl} \quad (23)$$

As previously noted, winch transport is conducted in underground tunnels with gradients ranging from 0° to 45°, encompassing horizontal, mildly inclined, and steeply sloped routes. These tunnels connect two vertically offset levels, separated by a height H_u (Figure 4). Knowing the value of H_u allows one to either assume or calculate the total length of the inclined tunnel, i.e., the transport distance L_u based on Equation (24), or conversely, to determine the achievable vertical difference H_u (25) or the tunnel inclination angle α_u (26). These relatively simple trigonometric relationships make it possible to define the geometric parameters of the transport path, which subsequently enable the calculation of the winch loading conditions and the appropriate selection of drive motor power. In the case of horizontal tunnels with an inclination of 0 degrees, Equations (24)–(26) do not apply, as the force values are not dependent on the inclination angle α_a. This issue arises only when $\alpha_a > 0$.

$$L_u = \frac{H_u}{\sin(\alpha_u)} \quad (24)$$

$$H_u = L_u \cdot \sin(\alpha_u) \quad (25)$$

$$\alpha_u = \arcsin\left(\frac{H_u}{L_u}\right) \quad (26)$$

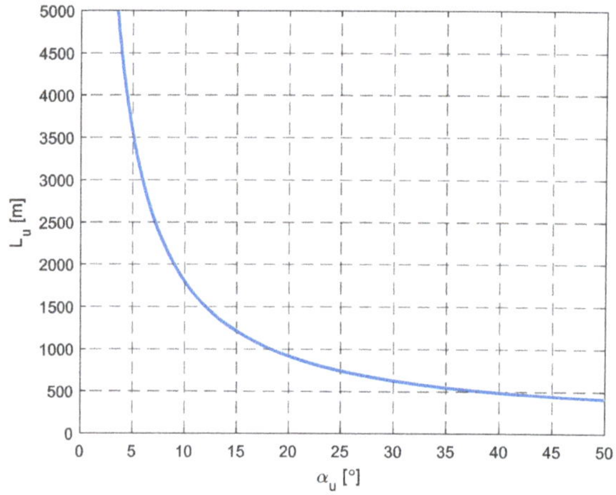

Figure 4. The transport distance L_u in the function of tunnel inclination angle α_u for H_u = 300 m.

Figure 4 illustrates the variation in the transport distance L_u as a function of the tunnel inclination angle α_u for a vertical difference of H_u = 300 m. This information is critical for users who must construct and equip such tunnels, and, most importantly, achieve the specified transport capacity. It should be noted, as mentioned previously, that these considerations apply only to tunnels where the tunnel inclination angle α_u does not exceed 45°. In practice, tunnels are most commonly encountered with tunnel inclination angles ranging between 10° and 20°. For instance, in this case, when α_u is 10°, the transport distance L_u is 1728 m, whereas for α_u at 20°, L_u is 877 m. The difference in length between these two scenarios is therefore 851 m.

It is evident that increasing the inclination of the tunnel leads to a significant shortening of its length; however, this also results in higher power consumption by the winch motor, assuming that the transport parameters (transported weight, time of a single cycle, and total cycle time) remain unchanged. On the other hand, the notable reduction in the tunnel's length considerably lowers the capital investment associated with its construction and commissioning. In this specific case (H_u = 300 m), assuming a linear increase in costs, the investment expense for the shorter tunnel is reduced by a factor of approximately 1.97. Therefore, it becomes imperative to determine the extent of the increased power demand of the winch motor while preserving the same transport parameters.

Figure 5 shows the velocity profile (time study) of the transport weight v_{pj} during a single cycle. It is assumed that the movement of the transported weight in the first section occurs with a velocity of v_{pja} (uniformly accelerated motion), starting from zero and increasing to a steady velocity v_{pju} over the acceleration time t_{cp}. This is followed by uniform motion at velocity v_{pju} for time t_{cu}, and finally, during time t_{co}, the transport system decelerates and comes to a complete stop (uniformly decelerated motion with velocity v_{pjb}).

The total time of a single transport cycle t_c (27) is the sum of the acceleration time t_{cp}, the uniform motion time t_{cu}, and the deceleration time t_{co}. During the time t_{cp}, the transport system moves with uniform acceleration a, covering a distance of L_{up}. Once the constant travel velocity v_{pju} is reached, the system moves with uniform velocity for time t_{cu}, covering a distance of L_{uu}. Breaking of the system, with deceleration b, occurs over a distance of L_{uo} (29).

$$t_c = t_{cp} + t_{cu} + t_{co} \tag{27}$$

$$L_u = L_{up} + L_{uu} + L_{uo} = 0.5a \cdot t_{cp}^2 + v_{pju} \cdot t_{cu} + 0.5b \cdot t_{co}^2 \quad (28)$$

$$L_u = 0.5\left[a \cdot t_{cp}^2 + b \cdot t_{co}^2\right] + v_{pju} \cdot t_{cu} \quad (29)$$

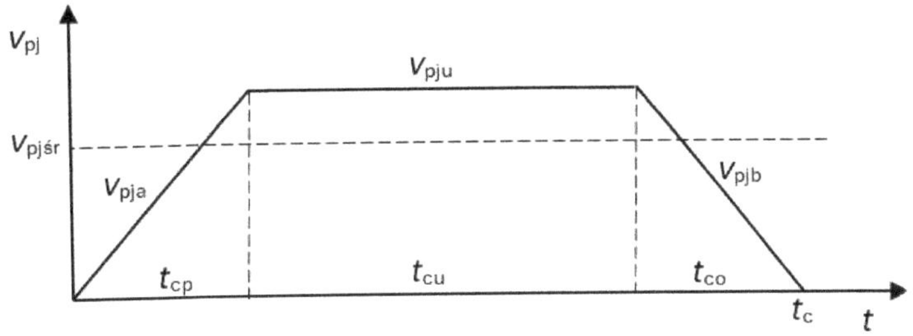

Figure 5. Graph of the transport velocity of the transported weight v_{pj} as a function of time t for a single cycle: v_{jpa}—velocity of the transported weight during acceleration to the steady velocity, v_{pju}—steady velocity, v_{jpb}—velocity of the transported weight during deceleration, $v_{pj\acute{s}r}$—average velocity of the transported weight, t_c—time of a single cycle, t_{cp}—acceleration time, t_{cu}—uniform motion time, t_{co}—deceleration time (braking time).

Assuming uniform motion along the entire transport distance L_u during the total cycle time t_c, the average transport velocity of the system v_{pjsr} can be determined using Equation (30). This is particularly useful during the stage of determining the tunnel inclination angle α_u (31), provided that mining and geological conditions, as well as the layout of the tunnels, allow for such optimization. In this context, the prospective user can select an appropriate value of α_u to minimize the capital investment required for the construction of the tunnel length L_u while still achieving the required transport capacity (v_{pj}, t_c).

In such a case, the transport capacity in the inclined tunnel using a winch system is measured by the number of transport cycles n_c that can be completed within a given time period T, i.e., the number of transport systems moved either upward or downward. The number of cycles n_c is then calculated as the ratio of the total available time T to the sum of the cycle time t_c and the preparation time t_{pzj} required to ready the transport system for travel (32). Accordingly, it is necessary to determine the required average transport velocity v_{pjsr} in order to complete the specified number of cycles n_c.

Knowing the winding force F_C acting on the transport system and the average transport velocity v_{pjsr}, the required winch motor power N can be easily determined using Equation (33). The winding force F_C for a given transport system is primarily dependent on the inclination of the tunnel, expressed by the tunnel inclination angle α_u, while the energy consumption of the transport process E_{ut} depends on the transport distance L_u and either the single-cycle time t_c (34) or a defined total operational time T (35). Thus, the energy consumption of the winch-driven transport system may be analyzed either per cycle t_c or over a specified time interval T.

$$v_{pjsr} = \left\{0.5\left[a \cdot t_{cp}^2 + b \cdot t_{co}^2\right] + v_{pju} \cdot t_{cu}\right\} \cdot t_c^{-1} \quad (30)$$

or

$$v_{pjsr} = \{0.5 v_{pju}(t_{cu} + t_c)\} \cdot t_c^{-1}$$

$$\alpha_u = \arcsin\left(\frac{H_u}{v_{pj\acute{s}r} \cdot t_c}\right) \quad (31)$$

$$n_c = \frac{T}{t_c + t_{pzj}} = \frac{T}{\frac{L_u}{v_{pjsr}} + t_{pzj}} = \frac{T \cdot v_{pjsr}}{L_u + t_{pzj} \cdot v_{pjsr}} \quad (32)$$

$$N = F_c \cdot v_{pjsr} \quad (33)$$

$$E_{utc} = t_c \cdot N \quad (34)$$

$$E_{ut} = T \cdot N \quad (35)$$

Figure 6 presents the run chart of the tunnel inclination angle α_u for H_u = 300 m and average transport velocities v_{pjsr} = 1.5, 2.0, and 2.5 m/s as a function of the single-cycle time t_c. Vertical lines indicate the values of t_c most commonly used by operators and recommended by winch manufacturers, while horizontal lines correspond to inclination angles α_u = 10° and 20°. Naturally, the higher the travel velocity v_{pjsr}, the shorter the achievable cycle time t_c, allowing for an increased number of cycles n_c for a given tunnel inclination angle α_u.

However, within the considered ranges of α_u (10°–20°), v_{pjsr} (1.5–2.5 m/s), and cycle times t_c (3, 6, 8, and 10 min), the values t_c = 8 or 10 min and possibly around 6 min can be taken forward for further analysis. This case is illustrated in Figure 7, where the selection of t_c depends on the assumed v_{pjsr}. In such scenarios, it may not be possible to achieve the required cycle time t_c unless the travel velocity v_{pjsr} is increased for a given inclination angle α_u. Therefore, average velocities of 2.0 and 2.5 m/s are recommended for further consideration. In both approaches, however, increasing the winch motor power is necessary.

Accordingly, for H_u = 300 m, the following technical parameters of the winch and the inclined tunnel are recommended for continued analysis:

- Tunnel inclination angle α_u = 10°–20°;
- Mean transport velocity v_{pjsr} = 2.0–2.5 m/s;
- Single-cycle time t_c = 8–10 min.

Under these conditions, the transport distance L_u will range from 877 m (α_u = 20°) to 1728 m (α_u = 10°). Ignoring investment costs at this stage, the next step in the analysis must involve determining the required winch drive power N and the energy consumption of the process E_{utc}.

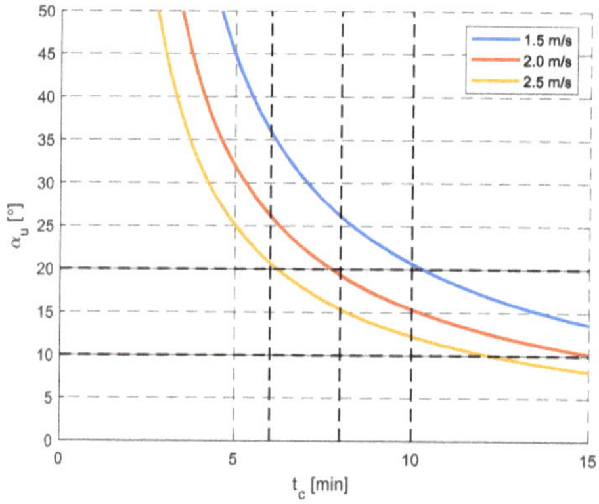

Figure 6. The tunnel inclination angle α_u, (max. 50°) for H_u = 300 m and v_{pjsr} = 1.5, 2.0, 2.5 m/s, as a function of time t_c.

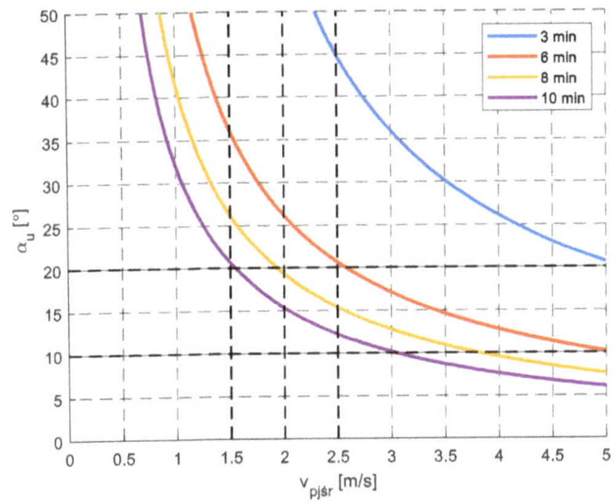

Figure 7. The tunnel inclination angle αu, for H_u = 300 m and t_c = 3, 6, 8, 10 min, as a function of mean transport velocity v_{pjsr}.

3. Computer Application

Further analysis of the technical parameters of the winch and the mine tunnel, in the context of selecting the winch motor power N and the energy consumption of the transportation process E_{utc}, requires the adoption of more detailed data. Utilizing the previously described analytical model and a proprietary computer application, relevant calculations were conducted for a selected transport system equipped with a rope winch installed in an underground inclined tunnel. These calculations aimed to determine the values of N and E_{utc} as functions of the tunnel inclination angle α_{u} = 10°–20°, transport velocity $v_{\mathrm{pjśr}}$ = 2.0–2.5 m/s, and the time of a single cycle t_c = 8–10 min (Figure 8). Additionally, the correctness of the selection of the return drum stroke of the rope tensioning mechanism was verified.

The following parameters were used for the calculations:

- Tunnel inclination angle, α_{u} = 10°–20° (10, 20°);
- Transported weight, G_m = 467 kN;
- Rolling resistance coefficient for G_m, μ_{pm} = 0.05;
- Mean transport velocity, $v_{\mathrm{pjśr}}$ = 2.0–2.5 m/s (2.0, 2.5 m/s);
- Rolling resistance coefficient for the unwinding rope, μ_{pl} = 0.05;
- Rolling resistance coefficient for the winding rope section from $\mathrm{K_{ZN}}$ to point A, μ_{pl1} = 0.05;
- Rolling resistance coefficient for the winding rope section from point A to the drive unit, μ_{pl2} = 0.05;
- Time of one cycle t_{c} = 8–10 min (8, 10 min).

For this purpose, a preliminary rope diameter of d = 42 mm was assumed, corresponding to a transported weight G_m = 467 kN and a unit rope weight q_l = 0.08 kN/m. Additionally, an initial rope tension force F_{nwl} = 20 kN and a stroke of the return drum ΔF = 1000 mm were adopted. As a result of the calculations (Figure 8), the tension force in the winding rope was determined to be F_C = 189.35 kN, while the tension force in the unwinding rope was F_S = 38.18 kN. These values confirm the adequacy of the selected rope parameters as well as the assumed initial rope tension force F_{nwl}.

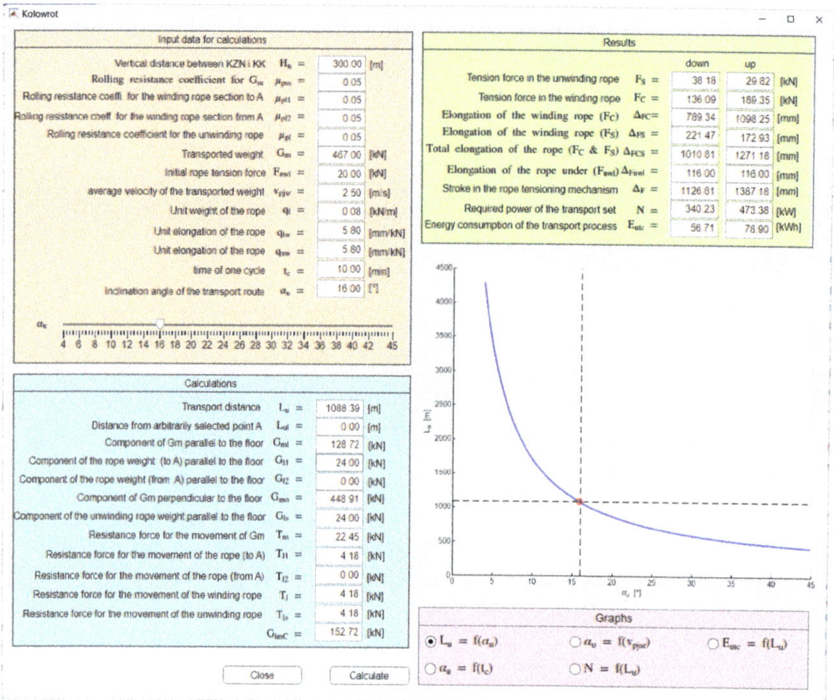

Figure 8. Main application window.

However, the stroke of the return drum in the rope tensioning mechanism was found to be ΔF = 1126.81 mm during descent and ΔF = 1387.18 mm during ascent. This indicates an increase of 387.18 mm above the initially assumed value, necessitating the modernization of the physical installation. Naturally, similar calculations can be performed with different input parameters to verify whether the actual performance aligns with the computed outcomes.

The data adopted for the calculations also enabled, as previously mentioned, the determination of the winch drive power and the energy consumption of the transport process. The results of these computations are shown in Figures 9 and 10. As expected, both the power demand and energy consumption of the transport process decrease with increasing transport distance L_u. A similar trend is observed with respect to t_c and v_{pjsr}. It is evident that doubling the transport distance L_u leads to an approximate 50% increase in power consumption. Likewise, energy consumption follows the same pattern. Increasing the transport velocity v_{pjsr} from 2.0 to 2.5 m/s results in a roughly 25% increase in power demand, with energy consumption exhibiting a comparable trend.

Ultimately, the decisive parameter for selecting both L_u and v_{pjsr} and thus for determining the required winch motor power N should be the single-cycle time t_c, which directly defines transport efficiency, as expressed by the required number of cycles n_c.

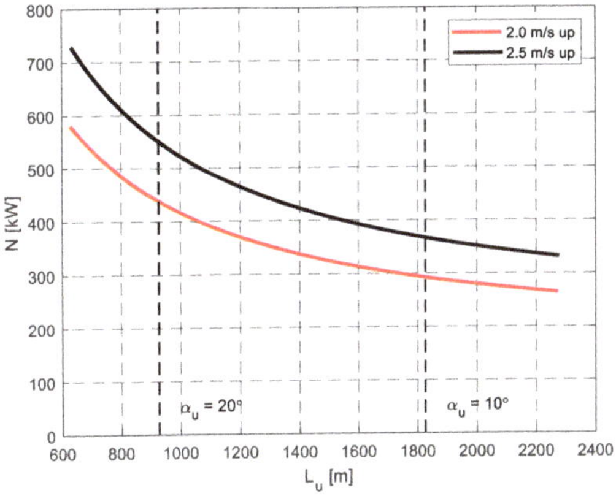

Figure 9. The necessary power N of the transport system as a function of L_u and v_{pjsr}.

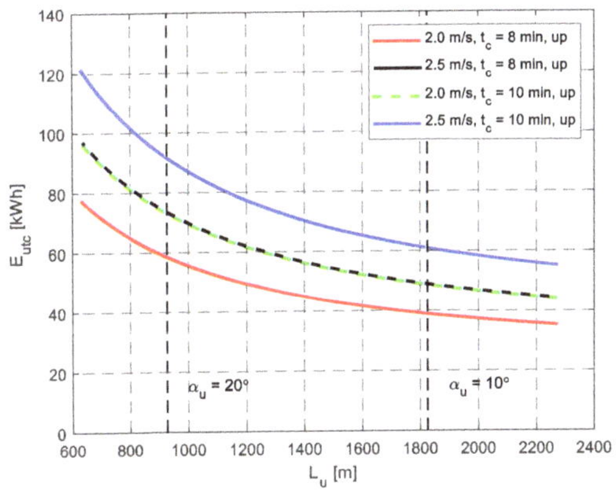

Figure 10. Energy consumption of the transport process E_{utc} as a function of L_u, and v_{pjsr} for t_c = 8 and 10 min.

4. Conclusions

The analytical model of the rope winch transport system, as previously mentioned, serves as a supporting tool in the selection process of the winch and other system parameters under specific conditions determined by the site of application. Therefore, the prospective user must be familiar with and able to specify the technical requirements for the winch, rope, transport distance, and safety devices.

In the context of winch determination of its technical parameters such as transport velocity, transported weight, rope tension force, drive power, and stroke of the return drum, the analytical model presented earlier proves to be particularly useful. Hence, when considering rail transport systems with rope winches, it is recommended that the selection procedure begins with defining the required transport capacity, i.e., the target number of operational cycles.

This allows for the determination of the necessary transport time and mean transport velocity. Subsequently, taking into account the existing network of tunnels and the vertical elevation difference, the user can determine and select the most advantageous tunnel inclination angle and, consequently, its length.

Author Contributions: Conceptualization, K.K.; Methodology, K.K.; Software, R.K.; Validation, R.K.; Formal analysis, K.K., R.K. and K.M.; Investigation, K.K. and T.W.; Data curation, R.K.;

Writing—original draft, K.K., R.K., K.M. and T.W.; Writing—review & editing, K.K. and K.M.; Visualization, R.K. and T.W.; Supervision, K.K.; Project administration, K.M.; Funding acquisition, K.K. and R.K. All authors have read and agreed to the published version of the manuscript.

Funding: This research was funded by AGH University of Krakow.

Data Availability Statement: The original contributions presented in the study are included in the article, further inquiries can be directed to the corresponding author.

Conflicts of Interest: The authors declare no conflict of interest.

References

1. Samset, I. *Winch and Cable Systems*; Springer Science + Business Media, B.V.: Dordrecht, The Netherlands, 1985.
2. Chen, J.; Zeng, B.; Xu, W.; Hu, S.; Wang, S.; Wang, K.; Zhang, W.; Wu, S.; Song, Z. Mechanical behaviour and damage constitutive model of semi-circular arch tunnels with straight walls under soaking conditions. *Eng. Fail. Anal.* **2025**, *169*, 109137. [CrossRef]
3. Haddad, J.S.; Denyshchenko, O.; Kolosov, D.; Bartashevskyi, S.; Rastsvietaiev, V.; Cherniaiev, O. Reducing Wear of the Mine Ropeways Components Basing Upon the Studies of Their Contact Interaction. *Arch. Min. Sci.* **2021**, *66*, 479–594. [CrossRef]
4. Dumitrescu, I.; Cozma, B.Z.; Itu, V. Study the mining winches using the CAD/CAE software. In Proceedings of the 16th International Multidisciplinary Scientific GeoConference SGEM, Sofia, Bulgaria, 30 June–6 July 2016.
5. Mushiri, T.; Jirivengwa, M.; Mbohwa, C. Design of a Hoisting System for a Small Scale Mine. *Procedia Manuf.* **2017**, *8*, 738–745. [CrossRef]
6. Ding, X.; Shen, G.; Tang, Y.; Li, X. Axial vibration suppression of wire-ropes and container in double-rope mining hoists with adaptive robust boundary control. *Mechatronics* **2022**, *85*, 102817. [CrossRef]
7. Park, S.; Choi, Y.; Park, H. Optimization of truck-loader haulage systems in an underground mine using simulation methods. *Geosyst. Eng.* **2016**, *19*, 222–231. [CrossRef]
8. Bharathan, B.; Sasmito, A.P.; Ghoreishi-Madiseh, S.A. Analysis of energy consumption and carbon footprint from underground haulage with different power sources in typical Canadian mines. *J. Clean. Prod.* **2017**, *166*, 21–31. [CrossRef]
9. Yu, K.A.; Yang, T.; Gong, C.Z. Design and Research of Double-Drum Winch with Anti-Pressure Wheel. *Adv. Mater. Res.* **2012**, *479–481*, 1709–1713. [CrossRef]
10. Shafi, M.S.; Lu, J.; Song, Z.; Fu, Y. Design and tension control of double drum Winch system for deep-sea exploration equipment. Advances in Engineering Research. In Proceedings of the 5th International Conference on Information Engineering for Mechanics and Materials 2015, Hohhot, China, 25–36 July 2015; pp. 1474–1480. [CrossRef]
11. Duong, T.G.; Nguyen, V.T.; Nguyen, T.D. Optimizing the weight of the two-level gear train in the personal rescue winch. *Arch. Mech. Eng.* **2021**, *68*, 271–286. [CrossRef]
12. Lee, J.H.; Yun, Y.W.; Hong, H.W.; Park, M.K. Tension control of wire rope in winch spooler using magneto rheological brake. *Int. J. Precis. Eng. Manuf.* **2016**, *17*, 157–162. [CrossRef]
13. Klich, A.; Kozieł, A. *Transport Szybowy*; Instytut Techniki Górniczej KOMAG: Gliwice, Poland, 2009. (In Polish)
14. Kevac, L.; Filipovic, M.; Rakic, A. Dynamics of the process of the rope winding (unwinding) on the winch. *Appl. Math. Model.* **2017**, *48*, 821–843. [CrossRef]
15. Yoo, S.; Kim, T.; Seo, M.; Oh, J.; Kim, J.; Kim, H.S.; Seo, T.W. Modeling and verification of multi-winding rope winch for facade operation. *Mech. Mach. Theory* **2021**, *155*, 104105. [CrossRef]
16. Boiko, A.; Naidenko, E.; Besarab, O.; Bondar, O. Analysis of Factors Affecting the Energy Efficiency of an Elevator Winch. In *Advanced Manufacturing Processes V, Proceedings of the 5th Grabchenko's International Conference on Advanced Manufacturing Processes InterPartner 2023, Odessa, Ukraine, 5–8 September 2023*; Lecture Notes in Mechanical Engineering; Tonkonogyi, V., Ivanov, V., Trojanowska, J., Oborskyi, G., Pavlenko, I., Eds.; Springer: Cham, Switzerland, 2023. [CrossRef]
17. Shi, Q.; Wang, D.; Zeng, X.; Guo, Y.; Wang, H.; Xu, Z.; Deng, Q.; Zhong, H.; Wang, B.; Li, C. Research on the Design of Multi-Rope Friction Hoisting System of Vertical Shaft Gravity Energy Storage System. *Appl. Sci.* **2024**, *14*, 7556. [CrossRef]
18. Chen, W.; Zhu, S.; Yue, R. Design of Control Device for Mine Endless Rope Winch. In *Advances in Engineering Research, Proceedings of the International Conference on Chemical, Material and Food Engineering, Kunming, Yunnan, China, 25–26 July 2015*; Atlantis Press: Dordrecht, The Netherlands; pp. 628–631. [CrossRef]

energies

MDPI

Article

The Technical and Economic Aspects of Using DC or AC Motors in the Drive of Hoisting Machines

Tomasz Siostrzonek *, Jacek Pytel and Tomasz Karpiel

Department of Power Electronics and Automation of Energy Transformation Systems, AGH University of Krakow, 30-059 Krakow, Poland; jpytel@agh.edu.pl (J.P.); karpieltomasz@gmail.com (T.K.)
* Correspondence: tsios@agh.edu.pl

Abstract: A hoisting machine is one of the main components of an underground mine process line. The drive systems currently used in these machines are based on direct current, synchronous, or induction (AC) motors. Power electronic converters are used to power and control the speed of the motor, the type and function of which depend on the type of motor. Choosing the right motor is the most important part of the design process. The decision must be based on an evaluation of technical and economic parameters. This article describes the most popular types of hoisting machine drive systems. Four motors that could potentially be used to build a hoisting machine with the assumed parameters were analysed. The selection of the motor was made based on the analyses carried out, and the implementation of the investment and commissioning of the machine with the selected motor confirmed the validity of this choice. The technical–economic analysis of the motor selection for this hoisting machine made it possible to determine how to choose a motor for a specific shaft hoist. The information contained in this article can serve as a guideline during in-depth analyses of motor selection for modernised or constructed hoisting machines due to the possibility of generalising its conclusions.

Keywords: hoisting machines; induction motor; synchronous motor; DC motor; efficiency of motors; economic analysis of efficiency; sensitivity analysis

1. Introduction

The mine shaft hoist is one of the most essential pieces of equipment in mining plants. It is the transport link between the surface and underground parts of the mine. Concerning the type of rope drive (the element that changes the rotational movement into the progressive movement of the vessel in the shaft), hoisting machines are divided into the following categories [1]:

- machines with a thrust wheel (Koepe pulley), where the hoisting rope is slung over the Koepe pulley and the movement of the skips/cages hooked to the ends of the rope is achieved by a suitable frictional coupling between the rope and the rope drive;
- machines with a drive drum, where the rope is wound on the drum when the hoisting vessel is raised and unwound when lowered.

Both types of arrangement are shown in Figure 1.

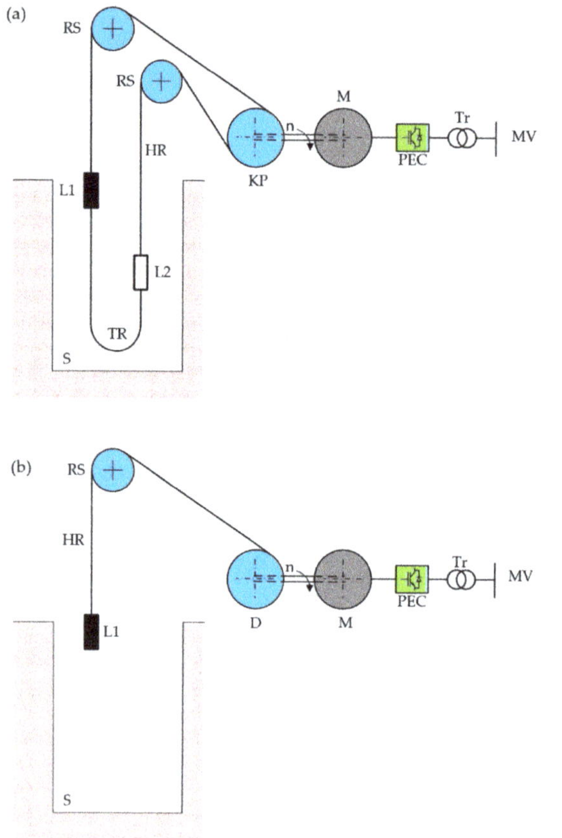

M – electrical motor; KP – Koepe pulley; D – drum; RS – rope sheaves/repel sheaves; L1 – load: skip/cage; L2 – skip/counterweight; HR – head ropes; TR – tail ropes; n – rotational speed; PEC – power electronics converter; Tr – transformer; MV – middle voltage supply; S – shaft

Figure 1. The scheme of the hoisting machine: (**a**) with the Koepe pulley, (**b**) with the drum.

The arrangement with the Koepe pulley (Figure 1a) has the possibility of building multi-rope systems to increase the safety of the shaft hoist operation. The disadvantage is the need for non-destructive testing of the hoisting ropes due to the inability to cut the rope for testing. In addition, in order to balance the forces in the ropes, so-called compensating ropes are used standing and suspended under both vessels. This ensures that the hoist is fully balanced and only the drive can be selected due to the static overweight resulting from the weight of the material/probe/people to be transported.

In addition to the mechanical parts of the drive system, the primary function in this system is the motor and the power electronics used to power and control the drive's speed. The type of linkage does not matter when selecting the motor. The motor's parameters are calculated based on the assumed load resulting from the driving diagram. The technical data of the converter and the converter transformer are determined based on the motor rating and the assumed operating conditions.

Due to its important function in the process line, the mine shaft hoist is counted among the basic equipment in mining plant operations. It consists of the following components [2]:

- hoisting machine;

- hoisting vessels (skips, cages);
- rope sheaves;
- rope attachments for compensation, guide and buffer ropes;
- carrier attachments for hoisting vessels;
- shaft signalling system and communication equipment.

When looking at a mine shaft hoist through the prism of the hoisting machine, the most important thing is the choice of drive motor. This determines the configuration of the hoisting machine's power (motor-current) circuit and the type of components used (converters, transformers, high-speed circuit breakers, etc.).

More than 90% of hoisting machines in Poland currently use DC motors of the PW series manufactured by DFME. A recent alternative to the DC motor is the alternating current motor (AC motor). This can be either a squirrel-cage induction motor or a synchronous motor. In the case of the former, a gearbox is required. In the case of a synchronous motor, it can be made as a slow-speed motor with a higher number of pole pairs. Regardless of the type of motor used, its power supply is provided by a dedicated static converter system of the appropriate type.

2. Arrangement with DC Motor

Drive systems for hoisting machines with DC motors are built as single- or twin-motor systems. Two-motor systems were used due to the lack of motors with sufficiently high technical parameters such as torque and power rating on the market. After the introduction of the PW 200 motor, the construction of twin-motor systems was no longer justified. It is also not true that twin-motor systems were built because of the possibility of increasing the machine's safety. Such systems were not designed to operate with a single motor. This may be evidenced by the inability to reconfigure the electrical circuit for single-motor operation. Single-motor operation also requires an appropriate correction of the calculations of the mine shaft hoist in the electrical part. The additional moment of inertia of the rotor of the non-working motor must be included in the calculations. This significantly increases the sum of the masses in motion of such a system and sometimes makes operation with a single-drive motor impossible.

2.1. Single-Motor System

Figure 2 shows a block diagram of a single-motor winding machine drive system with a converter system.

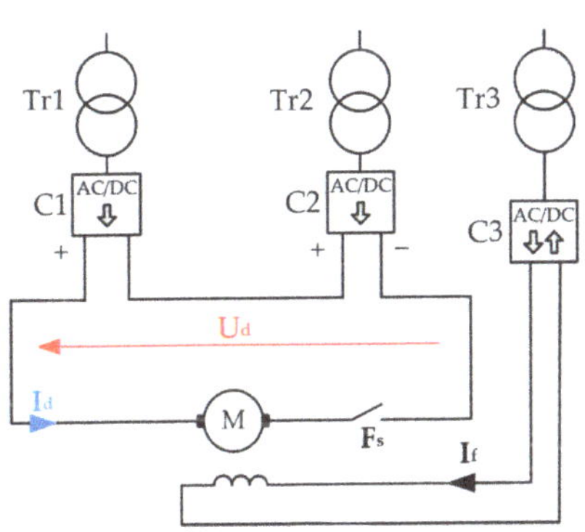

Figure 2. Diagram of a drive system with a separately excited DC motor and reversion through the excitation circuit.

The change of up/down direction is achieved by changing the direction of rotation of the motor, which is the result of a change in the direction of current flow in the excitation circuit. This is implemented through a converter marked C3. The investment cost of the entire system dictated this circuit configuration. The converters marked C1 and C2 are high-power converters with a rated current of more than 1 kA. The excitation converter (C3) is a much lower-power converter, i.e., from a few to several kilowatts, with a current rating of 100 to 250 A. To realise bi-directional energy flow, converters with oppositely connected semiconductor circuits are required, which increases the cost of building such a system. The cost of a converter increases with its power. However, the development of semiconductor components has meant that today, the price difference between single- and double-direction converters is insignificant. In addition, when changing the direction of rotation of the motor by changing the direction of current flow in the primary circuit, the control system of this set-up is simplified. Therefore, systems with so-called reversion through the primary circuit are increasingly common. A schematic of such a system is shown in Figure 3. C1 and C2 converters are characterised by bi-directional energy flow, as indicated by the arrows, in contrast to the C3 converter, where energy can only flow from the power source to the load—in this case, to the motor excitation circuit.

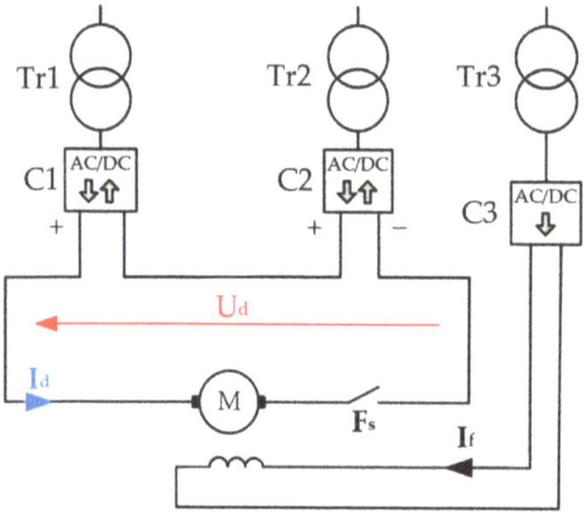

Figure 3. Diagram of a drive system with a separately excited DC motor and reversion through the armature circuit.

The systems shown in Figures 2 and 3 are 12-pulse systems, a typical arrangement used in winding machine drives. A detailed description of multi-pulse systems can be found in [3,4].

Converters C1 and C2 are supplied from separate converter transformers with the appropriate connection group, e.g., Dd0 and Dy11 (30° offset between the secondary side voltages of the transformers). The series connection of the converters ensures that the motor supply voltage level is adequate and the 12-pulse interaction of the drive with the main supply is realised. The ripple amplitude of the output voltage is significantly reduced, and the spectrum of the higher harmonics of the voltage supplying the converters is virtually devoid of harmonics 5 and 7. If one of the C1 or C2 converters fails, the system can operate as a 6-pulse system, i.e., one of the converters is disconnected, and the motor is powered only through the working system. The drive's speed is limited to half of the rated speed due to a 50% reduction in the supply voltage.

2.2. Twin-Motor System

For very high-power propulsion, twin-motor systems are used, the configuration of which originates in Pung's system (multi-machine). The use of this system was dictated by the lack of huge motors (above 3.6 MW). This arrangement allowed two motors with correspondingly lower power to be connected on a standard shaft and with a control system ensuring equal loading of both motors. Figure 4 shows a Pung's system with two PW series motors and converters.

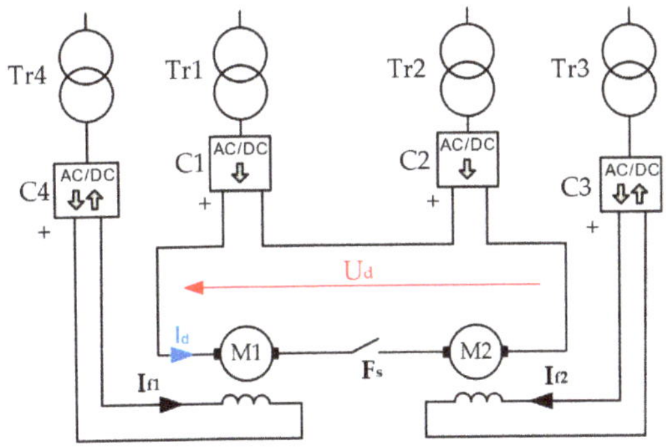

Figure 4. Two-motor system—Pung's system.

The construction of such a system is currently not economically and technically justifiable. Suppose that the propulsion power exceeds 6 MW (up to this power value, a PW 200 series motor can be used). In that case, an AC motor should be considered a more economically advantageous solution. The construction of a two-motor system is very costly due to the number of devices in this combination and the consequently greater complexity of the system, control, and protection. The consequence of this is significantly higher operating costs. The efficiency of the drive, which is calculated as a product of the efficiency of the individual components, also decreases. According to reliability theory [5], the greater the number of devices, the greater the probability of failures.

3. AC Motors Powered by Inverters

When considering the use of AC motors in a hoisting machine drive system, two types of AC motors can potentially be used: the induction motor (squirrel-cage or ring motor) and the synchronous motor [6–8]. The ring induction motor is not currently used due to its considerable cost and complex power supply concerning the operating parameters. Cage induction or slow-speed synchronous motors are used in newly built winders.

3.1. Squirrel-Cage Induction Motor

Squirrel-cage induction motors are manufactured as high-speed machines, i.e., their rotational speeds range between 500 and 3000 rpm. This rotational speed of the motor shaft is too high to make it possible to use it in the direct drive of a hoisting machine. Due to its design, it is not possible to construct a low-speed cage machine. Despite these disadvantages, squirrel-cage induction motors are used to drive winding machines thanks to a mechanical gearbox. The undoubted advantages of these motors are their production over an immense power and speed range and their durability.

The efficiency of induction motors currently available on the market reaches 96%, which is a considerable advantage compared with DC motors, which have efficiencies of the order of 90% [9–11].

3.2. Slow Synchronous Motor

A synchronous motor is an AC machine in which the rotor rotates at the same speed at which the field generated by the three-phase stator winding rotates. Unlike an induction motor, in which there is slippage which is a measure of the 'lagging' of the short-circuited rotor relative to the rotating field, in a synchronous motor, energy must be supplied to the excitation circuit (the rotating part connected to the motor shaft). In the case of hoisting machine drive motors, the excitation circuit's energy is provided from an external source, an AC/DC converter.

In the classic synchronous motor, the fundamental problem was the starting of such a motor. For this purpose, starting cages (asynchronous starting) or external starting machines were used. Such solutions are not feasible in the case of a hoisting machine drive due to the lack of smooth speed control and the problems of generating the correct value of electromagnetic torque. In this case, a system with a large number of pole pairs is used, which reduces the rotational speed, and by using converters, the motor is kept synchronised during start-up. This ensures that the correct drive torque is maintained in the speed range from zero up to the rated speed.

3.3. System with AC Motor and Direct Frequency Converter

The power supply for AC motors requires circuits that allow the infinitely variable speed control of the drive. Such systems built based on semiconductor components have led to a very rapid development of AC drive systems, successfully eliminating drives with DC motors. AC motors have also been successfully used in hoisting machines. In the case of AC motors, indirect or direct inverters are used to power and regulate their speed [7]. The two systems are fundamentally different due to the number of stages of variation of the supplied voltage.

Figure 5 shows squirrel-cage induction and synchronous motors powered by a direct frequency converter called a cycloconverter.

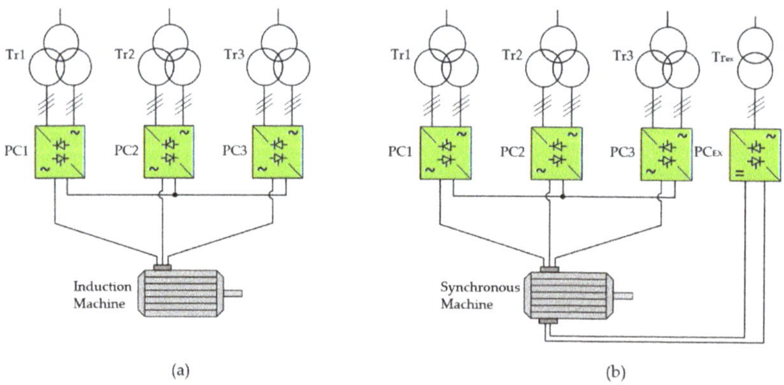

Figure 5. Cycloconverter drive with (**a**) induction motor and (**b**) synchronous motor.

The cycloconverter circuit consists of PC1, PC2, and PC3 converters in both cases. In the synchronous motor, a controlled rectifier PC_{EX} supplies the excitation circuit.

The operation of the cycloconverter is based on changing the frequency of the motor supply voltage, which has a direct effect on the smooth regulation of the motor speed according to Equation (1)

$$n_s = \frac{60 \cdot f_s}{p} \tag{1}$$

where

n_s—rotational speed of the field generated in the stator;

f_s—frequency of supply voltage (controlled by the cycloconverter);

p—number of pole pairs.

Thus, speed control in both directions is achieved over a wide range for generator and motor operation. Therefore, the mechanical characteristics of the drives in all four quadrants of the speed and torque axis system can be shaped. The system can operate as a motor or generator in any direction of rotation.

When using cycloconverters, the main problem is the output frequency of such a system. As standard, AC motors are built to be powered at 50 Hz. Typically, cycloconverters are built for an output voltage frequency of 15 Hz. This is due to the need to maintain a suitable ratio between the input and output frequency of the cycloconverter. Due to the modulation of the 50 Hz voltage waveform by the control signal, it is recommended that the output frequency should not be greater than one-third of the input frequency—in this case 50 Hz, resulting in a maximum output frequency value of 1.33 Hz.

Consequently, there is a natural reduction in motor power due to the significant reduction in voltage in proportion to the decrease in frequency. The loss of drive power relative to rated conditions forces an oversizing of the system, which is an obvious disadvantage of this solution. The second disadvantage of this system is its impact on the mains supply. This negative effect of the cycloconverter is particularly noticeable in terms of reactive power. Due to its design, the cycloconverter consumes reactive power, which poses a serious problem regarding charges imposed by the distribution system operator. The limiting parameter is tgφ = 0.4. When this is exceeded due to the operation of the equipment, penalties are imposed on the consumer as specified in the connection contract. For this reason, appliances must operate in such a regime where the limit value is not exceeded excessively.

3.4. System with AC Motor and Indirect Frequency Converter

An indirect frequency converter is a design in which there is a double change in the energy parameters transferred from the grid to the load or from the load to the source. Figure 6 shows block diagrams of drive systems with indirect frequency converters, squirrel-cage induction, and synchronous motors.

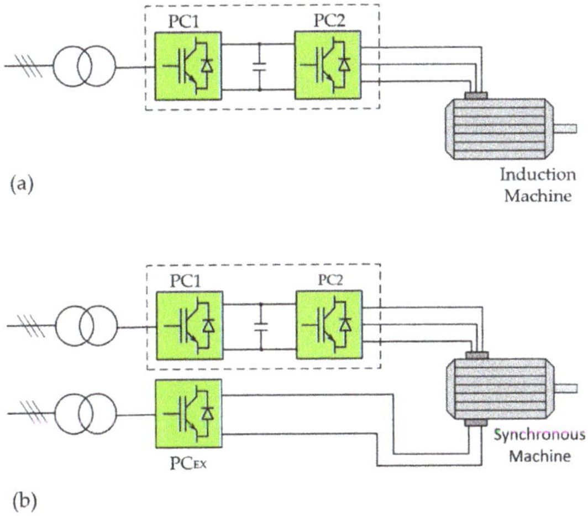

Figure 6. Indirect frequency converter with (**a**) induction motor and (**b**) synchronous motor.

Converters PC1 and PC2 are bi-directional AC/DC circuits that perform the rectifier (PC1) roles in the system and the inverter (PC2) in the case of motor operation of the drive system. The machine's generator operation is realised when PC2 operates as a rectifier and PC1 as an inverter.

The capacitor between the converters PC1 and PC2 constitutes the DC link circuit. Thanks to this system design, it is possible to reduce the system's negative impact on the mains through suitable control algorithms for the PC1 converter. So-called enhanced power factor (PFC) systems are currently used as PC1, which makes the entire drive system a purely resistive load for the mine's electrical network. There is no angular offset between the supply current and voltage indication. In the case of DC motors, the only way to reduce the impact on the mains was to use multi-pulse systems, which did not have the same measurable effect as PFC systems [12].

4. Analysis of the Technical Parameters of DC and AC Motors Used in Hoisting Machines

Table 1 summarises the basic parameters for the four motors. These motors meet the criteria defined for the selected design process. The primary criterion was the motor's rated torque, which could not be less than 480 kNm.

Table 1. Summary of technical parameters of motors.

	A	B	C	D
	DC motor	DC motor	AC synchronous MV motor	AC synchronous LV motor
Type	PW 202	PW 115	1DM5643-6XT08	AMZ 1600_16
Rated power P_N [kW]	3600	2350	3000	2600
Rated speed n [rpm]	45.86	45.86	36	46
Rated torque M_N [kNm]	750	489	795	540
Maximum torque M_{max} [kNm]	1600	980	1593	1350
Moment of inertia J [kgm^2]	74,000	35,500	28,500	14,000
Mass m [Mg]	77.2	57.45	54	46
Efficiency η [%]	92	89.9	93.1	93.5

For comparison, motors used in the hoisting machinery drive were selected, and the selection procedure was carried out for a specifically selected mining shaft hoist. Both DC and AC motors were analysed. Mine crews' preference for the motor type was ignored during the analysis. DC motors were analysed because of their production in Poland and high popularity among designers and mine employees.

During the design process, the motor is selected for its operating parameters—power and torque on the motor shaft. These two parameters result from the mass transported in the hoisting vessel and from the design of the mine shaft hoist itself, i.e., the depth of the shaft, the mass of the vessels, wheels, and ropes, i.e., generally, the masses in motion. Let us analyse the masses in progressive and rotational motion. Ideally, the rotor mass, and therefore the rotor moment of inertia of the hoisting motor rotor, should be as small as possible [13–16].

Evaluating the different motor types regarding technical parameters is essential for assessing investment costs (CAPEX) and determining their subsequent operation (OPEX) [17,18].

To select a hoisting motor combining technical and economic data, the equation proposed by the authors can be used:

$$W_{opt} = \frac{\frac{M_{max}}{M_N} \cdot \eta \cdot h_k \cdot P_N}{C^2 \cdot J \cdot m} \tag{2}$$

where

M_{max}—maximum torque [kNm];

M_N—rated torque [kNm];

h_k—number of hours needed for maintenance (brush changes);

P_N—rated power [kW];

c—price [mln PLN];

J—rotor's moment of inertia [kgm^2];

m—mass of motor [kg];

η—efficiency [%].

Thus, analysing the results in Figure 7, it can be concluded that separately excited DC motors, marked A and B, have significantly worse parameters than C and D motors. The rated parameters of torque, moment of inertia, maximum torque, and the power of individual motors were converted per 1 kW of power or per 1 kg of motor mass. This approach makes it possible to compare the motors despite the non-significant differences in the rated parameters.

3a	M_N/P_N [kN/kW]	max
A	0.21	
B	0.21	
C	0.265	X
D	0.21	

3d	M_N/m [kN/kg]	max
A	0.0097	
B	0.0085	
C	0.015	X
D	0.012	

3b	M_{max}/P_N [kN/kW]	max
A	0.44	
B	0.41	
C	0.531	X
D	0.519	

3e	M_{max}/m [kN/kg]	max
A	0.021	
B	0.017	
C	0.029	X
D	0.029	X

3c	J/P_N [kgm^2/kW]	min
A	20.6	
B	15.1	
C	9.5	
D	5.38	X

3f	J/m [kgm^2/kg]	min
A	0.96	
B	0.618	
C	0.53	
D	0.3	X

3g	P_N/m [kW/kg]	max
A	0.047	
B	0.041	
C	0.056	
D	0.057	X

Figure 7. Evaluation of individual relative parameters.

By analysing the individual parameters, it can be concluded that the D motor is the right choice. The particular technical parameters have an impact on the subsequent operation of the motor. Thus, a lower moment of inertia means that less energy is required to accelerate and decelerate the mass. For example, the choice of a motor with a lower moment of inertia has a further impact when carrying out system upgrades in Poland. The regulations in force do not allow for increasing the masses in motion in retrofitted systems. This means that when carrying out a system modernisation aimed at increasing the mining capacity of a plant, it is necessary to reduce the masses in motion, i.e., to transfer the mass from components such as rope wheels, motor rotors, or rope drives to the useful mass transported in the vessel.

One of the machines with a separately excited DC motor was analysed. According to the documentation of this mining shaft hoist, the assumed capacity was 40 skips/h. The skip load capacity was equal to 15 Mg. The transport of excavated material was not only coal but also waste rock. It is assumed that only 70% [19] of the extracted spoil is pure saleable coal. In this case, it was 10.5 Mg, which gives an hourly capacity of 420 Mg/h. The records from the hoist machine's operating recorder were analysed to check the actual hourly capacity of the hoist. The average extraction was 30 skips/hour, i.e., the hourly capacity was 315 Mg/h.

According to the daily operating schedule of the hoist [20], the time allocated to haulage (adequate working time) was two working shifts (2 × 8 h) five days a week. The remaining time was scheduled for all inspections, checks, and necessary planned and unplanned work (repairs, maintenance and overhaul) [21]. The time foreseen for replacing the electrographite brushes of the hoisting machine was 96 h/year, or an average of 2 h per week. The average weekly operating time of the hoisting machine is shown in Figure 8.

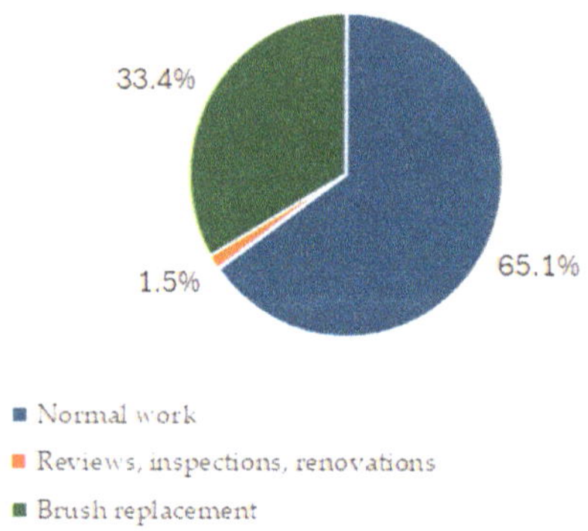

Figure 8. Weekly operating time of the hoisting machine.

5. Economic Analysis of the Use of DC and AC Motors in Hoisting Machines

Considering the technical findings in Section 4, a financial and sensitivity analysis was drawn up in a purpose-built model. To obtain comparable results for all analysed solutions, a fundamental assumption was made that ore availability for transport is fixed. The model was built based on fixed prices, and the discount rate was assumed to be 4%, a standard rate used in the industry for such analyses. In addition, the model assumed the following:

- a fair and equal useful life of 20 years for all motor types;

- an initial investment value that represents the total value of the motor, all accessories, and installation costs (CAPEX);
- discounted market salvage value [Residual Value], which refers to the estimated value of the motor at the end of its useful life.

5.1. Financial Analysis

When calculating the NPV indicator [17] for the hoisting machine motor, the part of the indicator relating to revenue remains the biggest problem. Attributing the total revenue of the mining plant to the hoisting machine is incorrect as it is only one part of the production sequence. Determining a percentage of all assets to estimate allocating a part of the total revenue will lead to different results. The reason for this is that there are significant differences in the size of the assets of the individual mines due to the technical solutions adopted to suit the different mining and geological conditions. For this reason, differential analysis was used to calculate the revenue part of the NPV indicator. The calculation of the total indicator is presented in Equation (3)

$$NPV = \sum_{t=1}^{n} \frac{(Revenue - OPEX)}{(1+r)^n} + \frac{Residual\ Value}{(1+r)^n} - CAPEX \tag{3}$$

where

r—discount rate;

n—lifetime.

In terms of the data making up the operating costs (OPEX) associated with the operation of the hoisting machinery, the difference is the cost of electricity consumed to transport the same amount of ore by machines with particular motors. Additional costs that differentiate the hoisting machine motors analysed are the costs associated with the maintenance of the equipment, consisting of the cost of human labour, the materials consumed, and the third-party services used which are necessary for periodic inspection and overhaul. The downtime induced by this work contributed to the lost revenue costs per year [LOSS], representing, for DC motors, between 37% and 38% of the discounted lifetime costs. These translate into a TCO [total cost of ownership] value representing the total discounted lifetime costs of the motors. The calculation of the indicator is given in Equation (4).

$$TCO = CAPEX + \sum_{t=1}^{n} \frac{(OPEX + LOSS)}{(1+r)^n} - \frac{Residual\ Value}{(1+r)^n} \tag{4}$$

In general, the TCO value [18] for AC motors represents, on average, 58% of the TCO generated by DC motors, which means that the total cost of ownership of AC motors is significantly lower, taking into account CAPEX and discounted OPEX, LESS, and Residual Value over the lifetime of the motors. This significant difference is mainly due to the presence of the lost revenue cost [LOSS] component in the calculation of the indicator for DC motors, which represents between 36.99% and 37.70% of the TCO indicator value, and the absence of this component in the calculation of the TCO indicator for AC motors. The next most weighted component on the TCO value is the amount of the discounted operating costs (OPEX). The calculated value of discounted operating costs (OPEX) for DC motors is 5.8 % to 9.4 % higher than those for AC motors. In the TCO structure, the discounted operating costs for DC motors were between 59.46% and 60.74% of the total indicator.

Due to the absence of the LOSS component in the TCO calculation for AC motors, its structure indicates a predominance of discounted operating costs (OPEX), representing between 95% and 96% of the TCO value. This means that this component has the most significant impact on the size of the total cost of ownership and should primarily be taken into account when selecting a motor. The discounted operating costs for DC motors are between 59.5% and 60.7% of the TCO structure. A significant share of the total cost of ownership indicator structure should lead to minimising operating costs and, thus, a lower TCO. In the analysis, the D motor achieved the lowest TCO value, which is the most desirable outcome and lower than the TCO of the C motor by 1.57%.

The share of CAPEX volume in the TCO calculation ranged from 2.36% to 4.79%, while the discounted Residual Value volume share ranged from 0.09% to 0.22%. These two components should not be significant when selecting a hoisting machine motor.

5.2. Sensitivity Analysis

A sensitivity analysis to a ±15% change in electricity prices resulted in a 53% to 62% change in TCO valuation in favour of AC motors. The sensitivity analysis to a ±15% change in coal prices only affects the LOSS components to which DC motors are sensitive. A 15% change in the price of coal resulted in a change in TCO from 5.5% to 5.8%.

A sensitivity analysis of the net present value of the investment (NPV) performed for changes in electricity prices and changes in coal prices (±15%) also confirmed that the recognition of reduced revenues resulting from the necessary downtime of the DC motors had the most significant impact on the indicator. Lower discounted operating costs for AC motors also translate into higher NPVs.

The difference between machine D and C represents 2.13% in favour of motor D. For the sensitivity analysis to a ±15% change in electricity prices, the difference in NPV between the AC motors reached a range of 2.80% to 1.67% in favour of motor D. The NPV sensitivity analysis to a ±15% change in coal prices also reached 3.29% to 1.57% in favour of motor D.

The magnitude differences between DC motors A and B in the NPV indicator are minimal. They are due to motor A's lower electricity consumption relative to motor B, whose discounted operating cost value more than compensates for motor B's slightly higher investment costs.

6. Discussion

Based on the analysis of technical parameters, it was concluded that the motor marked D was the best motor selected for the study. However, in addition to the technical analysis, it is necessary to determine the purchase and operating costs.

When assessing operating costs, one criterion is the maintenance and overhaul specified in the motor's technical and operating documentation. In addition to periodic maintenance, DC motors need to have their brushes changed and their commutator cleaned. This is in contrast to only periodic maintenance for AC motors.

Of course, there are organisational possibilities to move the brush replacement work to non-working days. Still, when planning an investment to build or replace a hoisting machine, it cannot be assumed that it will only work five days a week due to the need to increase the machine's effective operating time.

The calculated NPVs for all motors reached positive values. A higher NPV means a more profitable investment. The order of investment selection from highest NPV to lowest is as follows:

D > C > A > B

The calculated TCO values for DC motors are significantly higher than the TCO of AC motors, representing only 58% of their value on average. A lower TCO value means cheaper device use over its entire useful life cycle. The order of investment selection from lowest TCO values to highest is as follows:

D > C > A > B

For the two indicators adopted in the financial analysis, the best investment choice is AC motor D, followed by the second AC motor C. DC motors have lower net present values and a higher total cost of ownership.

From the analyses, it is clear that the D motor obtained the best results for the assumed needs, i.e., the specified technical parameters of the hoisting machine. This article does not state that this motor has the best techno-economic parameters of all the motors, but only for this one case. In each case, the choice of motor will require careful analysis from both the technical and economic sides.

7. Conclusions

The motor is the most critical component of the hoisting machine drive system. Its performance and operating parameters determine the entire mine shaft hoist's operational safety and efficiency. It should be noted that in most underground mines, two or three mine shaft hoists are available for transporting crew and ore. The hoisting machines are among the equipment which has a very long service life and are relatively rarely replaced due to technical wear and tear and the availability of new technology. It is, therefore, essential to decide on the type of motor prudently and to analyse both technical and economic aspects.

The analysis shows that the choice of AC motor (D motor) is justified from both technical and economic perspectives. In addition to the issues presented in the article, the choice of motor type determines the configuration of the entire drive system.

The reduction in investment costs is mainly due to the absence of additional devices to reduce the negative impact of the hoisting machine system on the mine's main supply.

Improvements in operating conditions are realised by the following:

- reduction in electricity demand;
- increasing the efficiency of the whole system (no additional components, e.g., reactive power compensators);
- elimination of machine elements requiring special supervision and maintenance (no commutator);
- no additional elements are subject to periodic checks and inspections, e.g., fast circuit breakers, reactive power compensators, and higher harmonics filters.

Unfortunately, little information can be found in the literature on selecting a motor for a winding machine and its potential benefits or losses. As can be concluded from this analysis, the choice of the motor determines the design of the entire electrical system. This translates into both investment and operating costs for the entire machine.

Author Contributions: Conceptualization, T.S.; Methodology, T.S.; Formal analysis, T.S. and J.P.; Investigation, T.S. and T.K.; Writing—original draft, T.S. and J.P.; Writing—review & editing, T.S.; Supervision, T.S. All authors have read and agreed to the published version of the manuscript.

Funding: This research received no external funding.

Data Availability Statement: The original contributions presented in this study are included in the article. Further inquiries can be directed to the corresponding author.

Conflicts of Interest: The authors declare no conflict of interest.

References

1. Szklarski, L.; Zarudzki, J. *Elektryczne Maszyny Wyciągowe. (Electric Hoisting Machines)*; Wydawnictwa Naukowe PWN: Warsaw, Poland; Kraków, Poland, 1998.

2. Rozporządzenie Rady Ministrów z dnia 30 Kwietnia 2004 r. w Sprawie Dopuszczania Wyrobów do Stosowania w Zakładach Górniczych (Regulation of the Council of Ministers of 30 April 2004 on the Authorisation of Products for Use in Mining Plants). Available online: https://isap.sejm.gov.pl/isap.nsf/DocDetails.xsp?id=WDU20040991003 (accessed on 4 January 2025).
3. Siostrzonek, T.; Chmielowiec, K.; Firlit, A.; Barczentewicz, S. Zalety i wady stosowania układów wielopulsowych w napędzie elektrycznym maszyn wyciągowych. *Przegląd Elektrotechniczny* **2020**, *5*, 96. [CrossRef]
4. Siostrzonek, T.; Chmielowiec, K.; Piątek, K.; Dutka, M.; Firlit, A. The use of multi-pulse systems in the power supply of hoisting machine drives to improve voltage parameters in mining plants. In Proceedings of the EPQU'20, 12th International Conference and Exhibition on Electrical Power Quality and Utilization, Krakow, Poland, 14–15 September 2020.
5. Lindstedt, P.; Sudakowski, T.; Grądzki, R. *Eksploatacyjna Niezawodność Maszyny i jej Teoretyczne Podstawy*; Wydawnictwo Instytutu Technicznego Wojsk Lotniczych: Warsaw, Poland, 2016.
6. Tiley, G.L. Effect of Mine Hoist Drives on the Electrical Power Supply System. *IEEE Trans. Ind. Appl.* **1987**, *IA-23*, 742–747. [CrossRef]
7. Noy, P.H. Advantages of PWM-Fed Induction Machine Applied to Mine Hoists. In *Hoist & Haul 2015, Proceedings of the International Conference on Hoisting and Haulage, Stockholm, Sweden, 21–25 June 2015*; Society for Mining, Metallurgy & Exploration: Englewood, CO, USA, 2015; pp. 141–149.
8. Stewart, A.T. AC Drive Selection for Mine Hoists. In *Hoist & Haul 2015, Proceedings of the International Conference on Hoisting and Haulage, Stockholm, Sweden, 21–25 June 2015*; Society for Mining, Metallurgy & Exploration: Englewood, CO, USA, 2015; pp. 191–202.
9. Figura, R.; Szychta, L.; Szychta, E. Wyznaczanie współczynnika sprawności silnika klatkowego pracującego ze zmienną prędkością obrotową. (The induction-cage motor efficiency determination for variable speed working). *Zesz. Probl.–Masz. Elektr.* **2005**, *73*, 163–168.
10. Repo, A.K.; Montonen, J.; Sizonenko, V.; Lindh, P.; Pyrhönen, J. Energy Efficiency of Hoisting Motors. In Proceedings of the International Conference on Electrical Machines ICEM'2014, Berlin, Germany, 2–5 September 2014.
11. Kumar, N.; Chelliah, T.R.; Thangaraj, R.; Srivastava, S.P. Economical Analysis of Induction Motor for a Mine Hoist Load Diagram. In Proceedings of the World Congress on Information and Communication Technologies, Mumbai, India, 11–14 December 2011.
12. Gąsiorski, A.; Posyłek, Z.; Wróbel, M. Aktywny korektor współczynnika mocy PFC w układach jednofazowych z prostownikiem dwupołówkowym. *Przegląd Elektrotechniczny* **2017**, *1*, 149–152. [CrossRef]
13. Johansson, P. Losses and Efficiency in Hoisting Systems. In *Hoist & Haul 2015, Proceedings of the International Conference on Hoisting and Haulage, Stockholm, Sweden, 21–25 June 2015*; Society for Mining, Metallurgy & Exploration: Englewood, CO, USA, 2015; pp. 163–181.
14. Emadi, A. *Energy-Efficient Electric Motors*; CRC: Boca Raton, FL, USA, 2004.
15. Krause, P.C. *Analysis of Electric Machinery*; McGraw-Hill: New York, NY, USA, 1986.
16. Jordan, H.E. *Energy Efficient Electric Motors and Their Application*; Van Nostrand Reinhold Company: New York, NY, USA, 1983.
17. Łada, M.; Kozarkiewicz, A. *Zarządzanie Wartością Projektów. Instrumenty Rachunkowości Zarządczej. (Managing the Value of Projects. Management Accounting Instruments)*; C.H. Beck: Warszawa, Poland, 2010.
18. Rogowski, W. *Rachunek Efektywności Inwestycji. (Investment Efficiency Account)*; Oficyna a Wolters Kluwer Business: Warszawa, Poland, 2013.
19. Osoba, M. Odkamienianie urobku surowego węgla kamiennego. (Decalcification of raw coal spoil). *Górnictwo Geol.* **2011**, *6*, 167–179.
20. Karpiel, T. Optimalisation of the Security Features Used in the Control System of the Winder for Use with the Winder of the Hydraulic Brake System. Ph.D. Thesis, AGH University, Krakow, Poland, 2020.
21. Rozporządzenie Ministra Energii z dnia 23 Listopada 2016r. w Sprawie Szczegółowych Wymagań Dotyczących Prowadzenia Ruchu Podziemnych Zakładów Górniczych (Regulation of the Minister of Energy of 23 November 2016 on Detailed Requirements for the Conduct of Underground Mining Operations). Available online: https://isap.sejm.gov.pl/isap.nsf/DocDetails.xsp?id=WDU20170001118 (accessed on 4 January 2025).

energies

MDPI

Article

Impact of Monthly Load Variability on the Energy Consumption of Twin Belt Conveyors in a Lignite Mine

Mirosław Bajda [1], Leszek Jurdziak [1,*] and Zbigniew Konieczka [2]

1 Faculty of Geoengineering, Mining and Geology, Wrocław University of Science and Technology, 50-370 Wrocław, Poland; miroslaw.bajda@pwr.edu.pl
2 PGE Górnictwo i Energetyka Konwencjonalna S.A, Oddział KWB Bełchatów, 97-400 Bełchatów, Poland
* Correspondence: leszek.jurdziak@pwr.edu.pl

Abstract: This study examines the influence of average monthly load variations on the unit energy consumption indicator (ZE) of belt conveyors, defined as the energy required to transport 1 Mg of bulk material (coal) over a distance of 1 km. The analysis is based on energy consumption data from twin belt conveyors operating in a Polish lignite mine over four years. Monthly records included transported mass, conveyor operating time, energy consumed, and atmospheric conditions (temperature and humidity). The results indicate statistically significant differences in ZE between the analyzed conveyors. A relative increase in average load by 35% (from 2000 Mg/h) corresponded to a 26% reduction in ZE, with a specific decrease of 90 Wh/Mg/km. The findings highlight that optimizing conveyor load distribution can significantly reduce energy consumption—by up to 30%. These insights underscore the potential for energy savings through better conveyor load management without requiring significant infrastructural investments.

Keywords: belt conveyor; energy consumption; unit energy consumption; conveyor load; theoretical capacity

1. Introduction—Collected Data

Belt conveyors are not only a more cost-effective alternative to wheeled transport [1–7], but they are also more environmentally friendly due to their use of electric motors for propulsion and bulk material handling [8–10]. Additionally, when operating on downhill slopes, they can serve as energy generators [11,12], contributing to energy recovery and efficiency improvements.

Beyond their primary function in material transport, belt conveyors present innovative opportunities for energy management. They can serve as a form of dry energy storage, similar to pumped-storage power plants [13], by utilizing gravitational energy. Additionally, they can help mitigate renewable energy curtailment by enabling the seasonal transfer of surplus energy from summer to winter. This can be achieved by integrating conveyors with bucket-wheel excavators that remove overburden and expose lignite seams near power plants, effectively creating coal storage reserves [14]. In some cases, conveyors can be powered directly by solar panels, providing both a renewable energy source and protection for transported materials against adverse weather conditions [15].

This study explores the impact of varying conveyor loads on energy consumption, analyzing extensive operational data collected from a Polish lignite mine over four years. The findings aim to enhance the efficiency of belt conveyor systems and identify potential strategies for reducing energy consumption in large-scale mining operations.

Energy and Economic Efficiency of Belt Conveyors

The energy and economic efficiency of belt conveyors, both in standard applications and specialized scenarios like those mentioned earlier, depend largely on the energy required to transport a unit of mass over a distance of 1 km. A unit energy consumption indicator (ZE) was introduced in previous studies [16–21] to facilitate accurate comparisons between different conveyor systems, providing a standardized metric for evaluating conveyor performance. This indicator also enables the optimal selection of conveyors for specific transport tasks, including open pit optimization [22–25] and the selection of belt conveyors for underground operations [26].

Extensive research, including both laboratory and industrial studies, has been conducted to analyze the primary sources of conveyor motion resistance. Findings indicate that indentation rolling resistance [27–31] contributes to over 60% of total motion resistance (Figure 1). Researchers have also sought to systematically assess the energy efficiency of belt conveyors [21], and explore methods for reducing energy consumption.

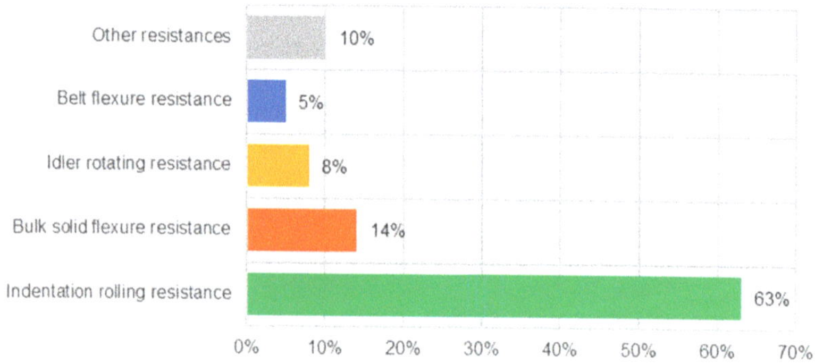

Figure 1. Shares of different components in the total motion resistance of a belt conveyor [9].

Key areas of investigation include:

- Reduction of idler resistance [32–35],
- The development of energy-saving belt conveyors and conveyor belts [36–39],
- Adjustments to belt speed to improve trough loading efficiency [8,37,40–42],
- Optimization of chute design to minimize energy losses [43–47].

A methodology for comparing the energy consumption of belt conveyors has been presented in previous studies [16]. These studies have dmonstrated that variations in monthly energy consumption are statistically significant, making direct comparisons beteween different conveyors challenging. Even when operating under nearly identical conditions, fluctuations in transported mass and monthly energy usage resulted in considerable differences.

This study focuses on two twin conveyors operating in parallel coal transport lines. Despite their nearly identical construction and function—both transferring coal from the same pit to the same power plant during similar periods—they exhibited negligible small differences in length (Table 1) and, more importantly, in the total mass of coal transported and the total time of their operation (Table 2). This difference was found to directly influence their average capacity and, consequently, the degree to which their theoretical capacity was utilized (Table 3).

Table 1. Technical data of Conveyors A and B.

Conveyor Construction Data	A	B
Length in m	1012.6	1018.5

Table 1. *Cont.*

Conveyor Construction Data	A	B
Belt speed in m/s	5.24	5.24
Theoretical mass capacity in Mg/h	6400	6400
Theoretical volume capacity Mg/h	8000	8000
Number of pulleys in a set	3	3
Idler length in mm	670	670
Idler diameter in mm	191	191
Idler set spacing in m	1.24	1.24
Trough angle	45	45
Belt type	St3150	St3150
Belt width in mm	1800	1800
Coal bulk density in Mg/m^3	0.8	0.8

Table 2. Collected operational data (January 2014–December 2017).

Collected Operating Data	A	B
Number of monthly measurements	36 *	48
Total mass transferred in Mg	42,096,610	46,540,032
Total operating time in hours	17,672.9	20,151.84
Total energy consumption in MWh	12,058.42	13,994.02
Average monthly operating time in hours	490.91	419.83
Standard deviation	60.98	91.28
Coefficient of variation in %	12.42%	21.74%
Minimum	352.1	198.6
Maximum	629.4	584.7
Range of change	277.3	386.1
The average amount of mass transferred in Mg	1,169,350	969,584
Standard deviation	198,752	243,846
Coefficient of variation	17.00%	25.15%
Minimum	760,503	424,723
Maximum	1,695,160	1,574,020
Range of change	934,657	1,149,297
Average energy consumption in MWh	334.956	291.542
Standard deviation	43.88	62.97
Coefficient of variation	13.10%	21.60%
Minimum	212.0	140.7
Maximum	441.6	406.5
Range of change	229.6	265.8

* During the periods January–July 2014 and June–October 2015, mass transfer data for Conveyor A was unavailable due to a scale failure.

Table 3. Calculated performance and ZE indicators.

Calculated Capacities and ZE Indicators	A	B
Average efficiency for total values in Mg/h	2381.99	2309.47
The degree to which the theoretical capacity was utilized in %	37.22%	36.09%
Unit indicator ZE for total energy consumption in Wh/Mg/km	286.45	300.69
Average monthly efficiency Q in Mg/h	2370.56	2296.89
Standard deviation	146.622	142.201
Coefficient of variation	6.19%	6.19%
Minimum	2119.17	2045.89
Maximum	2693.29	2692.01
Range of change	574.12	646.12
ZE indicator in Wh/Mg/km	288.57	303.58
Standard deviation	18.18	19.12
Coefficient of variation	6.30%	6.30%
Minimum	256.87	254.7
Maximum	336.35	344.57
Range of change	79.48	89.87

The width and speed of the belts were identical for both conveyors, ensuring that variations in energy consumption were not due to structural differences or external weather conditions. Instead, discrepancies in energy use were solely attributed to differences in transferred coal mass and achieved monthly performance levels.

Because these conveyors operated in parallel, direct comparison was complicated by significant monthly and overall differences in transported coal mass and operating time. Consequently, their actual average efficiency—measured as the ratio of transported mass to operating time—varied monthly. Additionally, Conveyor A's data were incomplete, missing 12 months of measurements.

Direct energy consumption values were not used to ensure a fair and reliable comparison. Instead, the unit energy consumption (ZE) index was analyzed, allowing for a standardized evaluation of conveyor efficiency across varying operational conditions.

2. Method of Comparison of Energy Consumption by Twin Conveyors

The energy consumption of two twin belt conveyors (A and B) operating in a lignite mine in Poland was compared over a four-year period, from January 2014 to December 2017. Both conveyors transported lignite from the mine to the power plant and had nearly identical specifications, including belt type and length, leading to their classification as "twins." Despite operating under similar conditions, differences in energy consumption were observed. The aim of this comparison was to determine whether the monthly variability in load impacted energy consumption.

The method of comparison focused on calculating the unit energy consumption (ZE) index required to transport 1 Mg of coal over a 1 km distance (Wh/Mg/km), which provided a consistent basis for comparison.

One important observation from the data was the differences in the coefficient of variation (CV) for various parameters. The CV for energy consumption was 13.1% for Conveyor A and 21.6% for Conveyor B, which was somewhat lower than the CV for mass transported (17.0% for A and 25.15% for B). This suggests that energy consumption was more stable compared with the variability in mass transported and operational hours. In contrast, capacity (measured in Mg/h) and the energy consumption (ZE) index per unit mass and distance showed much less fluctuation, with CVs of 6.19% and 6.30%, respectively, for both conveyors. This made these indices more reliable for comparison.

When analyzing the energy consumption indices, it was found that the unit energy consumption on Conveyor A averaged 288.57 Wh/Mg/km, while Conveyor B averaged 303.58 Wh/Mg/km, indicating a 5.23% higher energy consumption on Conveyor B. This difference was statistically significant. Despite these global differences, the monthly variation in unit energy consumption (ZE) index was considerable, ranging from 254.7 Wh/Mg/km to 344.6 Wh/Mg/km for Conveyor B and from 256.87 Wh/Mg/km to 336.35 Wh/Mg/km for Conveyor A.

To further explore these differences, statistical tests were conducted on the monthly data, revealing a negative correlation between the capacity of the conveyors and the energy consumption indicators (ZE). This suggests that higher capacity led to lower energy consumption, a relationship that was consistent across both conveyors. It was hypothesized that the variation in energy consumption could be influenced by operational factors such as monthly load fluctuations.

Ultimately, this analysis highlighted the fact that despite the conveyors' similar design, significant and persistent differences in energy consumption remained. Further investigation is needed to explore the underlying factors contributing to these differences, particularly those associated with operational (load variability) and weather conditions (temperature).

3. Results and Discussion

3.1. Comparison of the Average Monthly Capacities of the Tested Conveyors

During the analyzed period, the average capacity of Conveyor A was 2381.99 Mg/h, with an energy consumption index (ZE_A) of 286.45 Wh/Mg/km. Conveyor B had an average capacity of 2309.47 Mg/h, with a slightly higher energy consumption index (ZE_B) of 300.69 Wh/Mg/km (Table 3). This means Conveyor B consumed 14.24 Wh/Mg/km more energy per transported unit than Conveyor A. Although the relative difference of 4.97% was not substantial, it was statistically significant and was caused by a 1.13% increase in average conveyor capacity.

Interestingly, an inverse relationship was observed between capacity and energy efficiency. Despite being less loaded, Conveyor B had a lower average monthly capacity (72.51 Mg/h, or 3.04%, compared with Conveyor A). This suggests that higher unit energy consumption is associated with lower capacity utilization. A strong negative correlation was identified between ZE and conveyor capacity ($R^2 = -0.84$ and -0.85 for Conveyors A and B, respectively). A 35% relative increase in capacity (from 2000 up to 2700 Mg/h) led to a 26% reduction in ZE (from 340 down to 250 Wh/Mg/km), equating to a 90 Wh/Mg/km decrease. Proper load distribution could reduce unit energy consumption by up to 30%.

A statistical comparison of two independent sample distributions was conducted to verify the significance of these differences. The results, illustrated in Figure 2, show the histograms of monthly average capacities for both conveyors. The modal value for Conveyor A (~2400 Mg/h) is slightly higher than that for Conveyor B (~2300 Mg/h), indicating a small but measurable shift in performance.

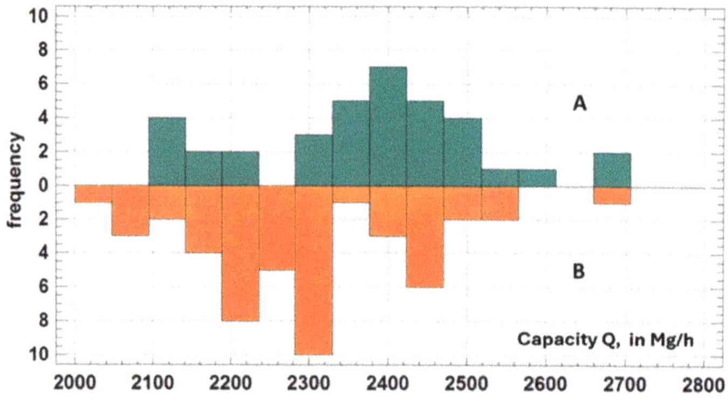

Figure 2. Comparison of the histograms of monthly average capacities (Q) for Conveyors A and B.

A more detailed view of the density distributions (Figure 3) confirms this shift. The peak value for Conveyor A is approximately 135 Mg/h higher than that for Conveyor B, which is almost twice the difference observed in the mean values (72.51 Mg/h). Additionally, the efficiency distribution of Conveyor A shows slight left-hand asymmetry, while Conveyor B exhibits a right-hand asymmetry, suggesting different operational characteristics.

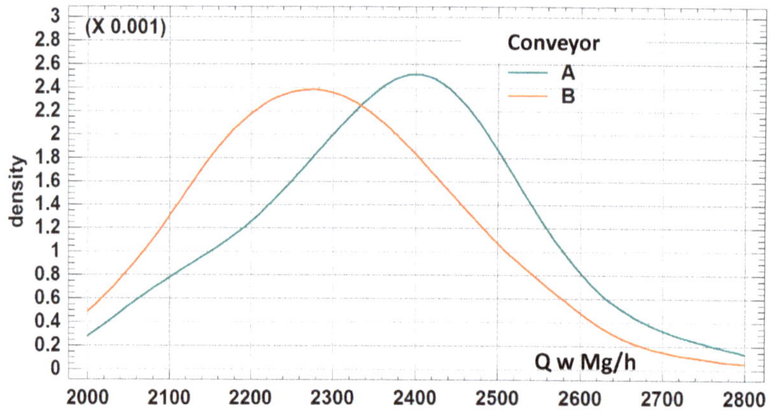

Figure 3. Density traces of monthly average capacities (Q) for Conveyors A and B.

3.2. Statistical Analysis of Conveyor Capacities

Table 1 presents key descriptive statistics for the analyzed datasets of monthly capacities. Notably, Conveyor A consistently exhibited higher values across various measures of central tendency (mean, median), with differences ranging from 3% to 4%. The median showed the most significant difference (−4.36%), while the harmonic mean differed the least (−3.1%).

In terms of dispersion, Conveyor A displayed more significant variability, with a higher standard deviation (146.62 Mg/h vs. 142.20 Mg/h) and variance. Conveyor B's variance was 5.94% lower, and its standard error was 16.01% lower, indicating slightly more consistent performance (Table 4).

Table 4. Basic statistics for monthly average capacities (Q) of Conveyors A and B.

Descriptive Statistics	Conveyor A	Conveyor B	Absolute Difference	Relative Difference (%)
No. of measurements	36	48	12	33.33
Mean in Mg/h	2370.56	2296.89	−73.67	−3.11
Median in Mg/h	2390.60	2286.37	−104.23	−4.36
Geometric mean in Mg/h	2366.14	2292.63	−73.51	−3.11
Harmonic mean in Mg/h	2361.69	2288.42	−73.27	−3.10
Standard deviation in Mg/h	146.62	142.20	−4.42	−3.02
Coefficient of variation	6.19%	6.19%	0.10	0.10
Standard error in Mg/h	24.437	20.525	−3.912	−16.01
Variance in Mg^2/h^2	21,498.1	20,221.2	−1276.9	−5.94
Minimum capacity in Mg/h	2119.17	2045.89	−73.28	−3.46
Maximum capacity in Mg/h	2693.29	2692.01	−1.28	−0.05
Range in Mg/h	574.12	646.12	72.00	12.54
Lower quartile (Q1) in Mg/h	2307.98	2200.30	−107.68	−4.67
Upper quartile (Q3) in Mg/h	2445.07	2409.83	−35.24	−1.44
Interquartile range in Mg/h	137.10	209.53	72.44	52.84
Skewness	0.0178	0.4232	0.4053	2273.16
Kurtosis	−0.0170	0.0559	0.0729	−427.76

The key observations are as follows:

- Consistent variation:
    - The coefficient of variation was nearly identical (6.19%) for both conveyors despite differences in mean values.
    - The maximum capacities were almost the same (~2693 Mg/h), while the minimum capacity was lower for Conveyor B.
- Higher dispersion in Conveyor B:
    - The range of fluctuations in Conveyor B's capacity was 72 Mg/h higher than in Conveyor A (646.12 Mg/h vs. 574.12 Mg/h).
    - This wider range may be due to seasonal and operational factors, requiring further investigation.
- Efficiency distribution differences:
    - Interquartile and intersextile ranges suggest a more concentrated distribution in Conveyor A, while Conveyor B exhibited greater dispersion at the extremes.
    - The kurtosis values suggest that both datasets approximate a normal distribution, though Conveyor A's slight negative kurtosis (−0.017) contrasts with Conveyor B's positive kurtosis (0.056).
    - Conveyor B exhibited stronger right-side skewness (0.42), indicating higher occurrences of lower-than-average capacities than Conveyor A (skewness = 0.017).

While statistically significant, the observed differences in capacities between the two conveyors do not appear to be solely due to construction differences. Instead, variations in load distribution, seasonal effects, and potential external factors likely contributed to the discrepancies. These factors warrant further exploration, particularly concerning their impact on energy consumption trends.

A separate study is needed to investigate atmospheric and seasonal influences on belt conveyor efficiency, as these may also explain the observed variability in capacity utilization.

Statistical Significance of Differences in Monthly Average Capacities

The differences observed visually (Figures 1 and 2) and identified through various descriptive statistics (Table 1) may appear significant but could still be coincidental. A series of tests was conducted to formally verify their statistical significance. Given that the differences affect not only central tendency measures but also positional statistics and density distribution traces (Figure 2), they are unlikely to occur by chance. However, a rigorous statistical approach is required.

Confidence Intervals for Monthly Average Capacities

To compare the means of the two conveyors, 95% confidence intervals were determined, as follows:

- Conveyor A: 2370.56 $\pm$ 49.61 [2320.95; 2420.17]
- Conveyor B: 2296.89 $\pm$ 41.29 [2255.59; 2338.18]

Additionally, a 95% confidence interval was calculated for the differences in the unit energy consumption (ZE) index, assuming equal variances:

- ZE difference: 73.67 $\pm$ 63.21 [10.47; 136.88]

Since this interval does not contain zero, we conclude that the difference in means is statistically significant at the 95% confidence level.

Student's *t*-test for Mean Comparison

A Student's *t*-test was performed to compare the means of the two samples. The null hypothesis (H0) assumed equal means, while the alternative hypothesis (H1) suggested a difference. Assuming equal variances, we obtain the following:

- t-statistic: 2.31884
- *p*-value: 0.02289

Since the *p*-value is less than 0.05, we reject the null hypothesis, confirming a statistically significant difference between the mean capacities of Conveyors A and B.

F-Test for Variance Comparison

Before performing the *t*-test, an F-test was conducted to check whether the two samples had equal variances. The standard deviations were as follows:

- Conveyor A: 146.622 [118.92; 191.26]
- Conveyor B: 142.201 [118.38; 178.18]
- Variance ratio: 1.0632
- 95% confidence interval for variance ratio: [0.57575; 2.0211]

The F statistic was 1.06315 with a *p*-value of 0.83486. As the *p*-value exceeds 0.05, we do not reject the null hypothesis, confirming no statistically significant difference in standard deviations.

Mann–Whitney (Wilcoxon) Test for Median Comparison

A Mann–Whitney test was conducted to compare the medians, as follows:

- Median capacities:
    - Conveyor A: 2390.6
    - Conveyor B: 2286.37
- Mean ranks:
    - Conveyor A: 49.6111
    - Conveyor B: 37.1667
- W-statistic: 608

- *p*-value: 0.02092

Since the *p*-value is less than 0.05, we reject the null hypothesis and confirm a statistically significant difference between the medians at the 95% confidence level.

Kolmogorov–Smirnov Test for Distribution Compatibility

The Kolmogorov–Smirnov (K-S) test was used to compare the distributions of both samples, as follows:

- DN Statistic: 0.3819
- K-S Statistic: 1.7323
- *p*-value: 0.00049

Since the *p*-value is below 0.05, we conclude that there is a statistically significant difference between the two distributions at a 95% confidence level.

Figure 4 compares box-and-whisker plots, scatter plots, and confidence intervals for the mean and median monthly capacities of both conveyors. The intervals do not overlap, reinforcing the statistical significance of the differences.

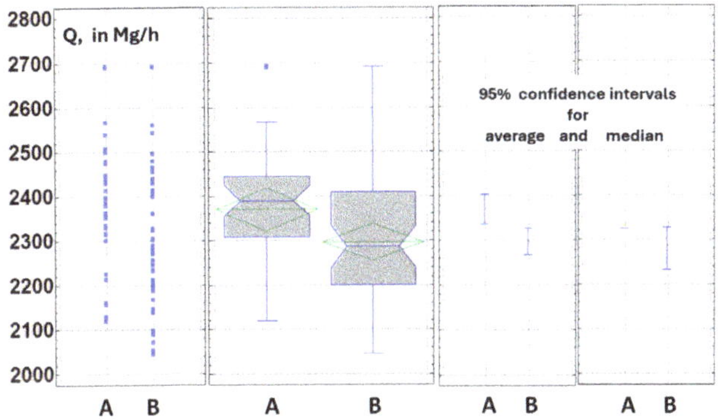

Figure 4. Graphical analysis of dispersion and confidence intervals of capacities Q for Conveyors A and B.

Figure 5 displays cumulative empirical distribution functions, showing that the distribution shift is most pronounced in the central region, with smaller variations at the tails.

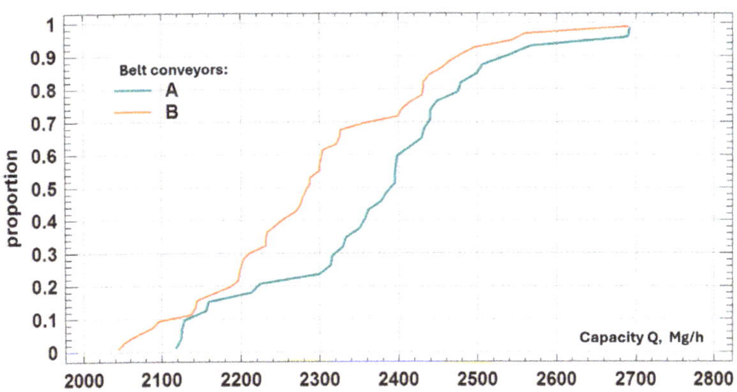

Figure 5. Chart of two cumulative empirical distributions for the monthly averages of capacity Q for Conveyors A and B.

Conclusion

The statistical tests confirm that the A and B conveyors' observed differences in monthly average capacities are significant. These differences primarily impact the variation of the unit energy consumption (ZE) index. The next chapter will explore the relationship between capacity and energy efficiency in greater detail through regression analysis.

3.3. Regression of the Unit Energy Consumption Indicator ZE Against the Average Capacity of the Conveyor

The impact of the monthly average capacity on the specific energy consumption index of the conveyor can be observed in the matrix graph (Figure 5), where all data points are plotted, differentiating between Conveyors A and B. The points are mixed and aligned along a negatively inclined diagonal, confirming a strong negative correlation (−0.84 and −0.85 for conveyors A and B, respectively).

The graphs (Figures 6 and 7) presents 84 complete data pairs (36 for Conveyor A and 48 for Conveyor B). More blue points (Conveyor A) are located in the lower part of the graph, while red points (Conveyor B) dominate the upper part. This indicates differences in the average specific energy consumption for Conveyor A (288.6 Wh/Mg/km) and Conveyor B (303.6 Wh/Mg/km), primarily due to differences in their monthly average capacities—2370.6 Mg/h for A and 2296.9 Mg/h for B.

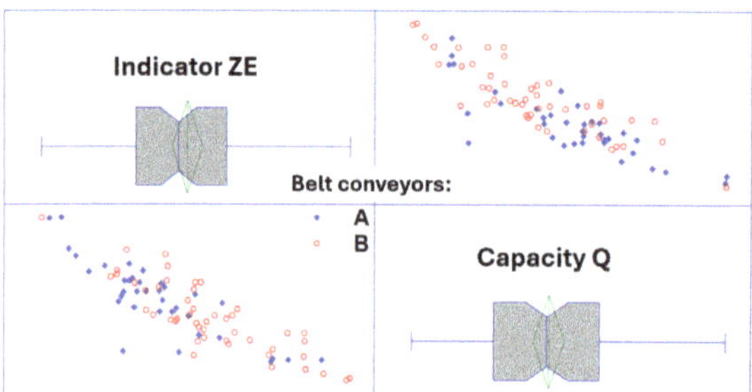

Figure 6. Matrix diagram showing the impact of monthly average capacity Q on the specific unit energy consumption indicator ZE for conveyors A and B.

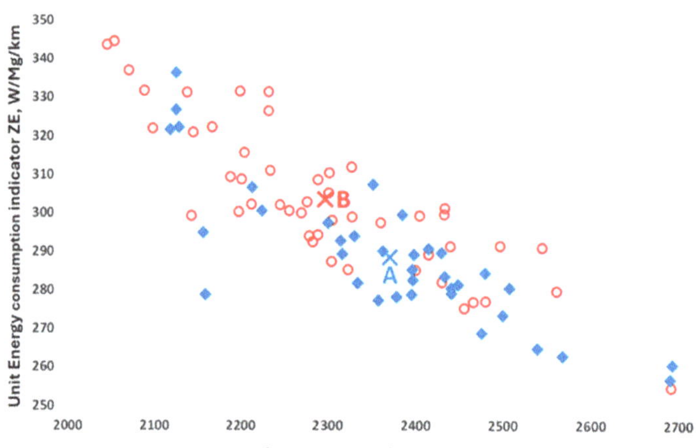

Figure 7. Scatter plot of average capacities Q and energy consumption indicators ZE for each month, with calculated averages (denoted by "X") for Conveyors A (blue diamonds) and B (red circle).

A simple regression test was performed to analyze the relationship between the dependent variable (unit energy consumption indicator ZE) and the independent variable (average efficiency Q) for Conveyors A and B separately and together.

For Conveyor A, the best-fit model was a hyperbolic function (inverse linear for Y, Figure 8), expressed as the following Equation (1):

$$ZE_A = 1/(0.0005512 + 0.000001235\ Q_A) \quad (1)$$

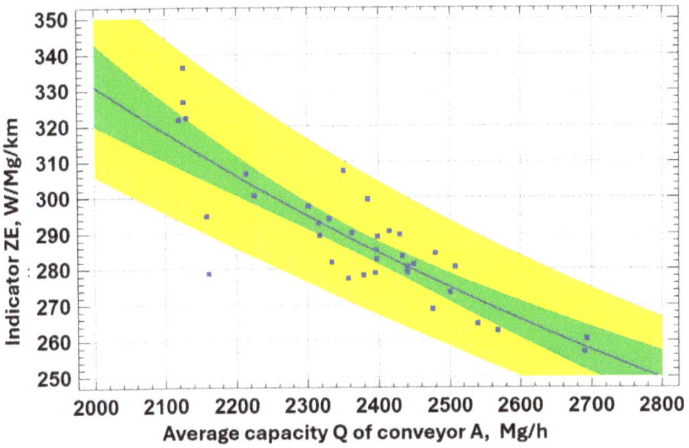

Figure 8. A simple regression of ZE for Conveyor A against average capacity Q over the analyzed months. Green color—2-sided 95% confidence intervals for the mean at given capacity Q. Yellow color—two sided prediction limits for new observations at the 95% level of confidence.

The estimation analysis, including ANOVA, yielded the following results:

- Correlation coefficient: 0.854862
- R-squared: 73.0789%
- R-squared (df corrected): 72.2871%
- Standard error of estimation: 0.0000796077
- Mean absolute error: 0.0000882741
- Durbin–Watson statistic: 0.909369 (p = 0.00001)
- Residual autocorrelation lag 1: 0.335697

Since the p-value in the ANOVA table (Table 5) is below 0.05, a statistically significant relationship exists between the indicator ZEA and capacity QA at a 95% confidence level.

Table 5. Coefficients and variance analysis for selected regression model for Conveyor A (D_f—degrees of freedom).

	Least Squares Method	Standard	Statistics		Value
Parameter	Estimate	Error	T		p
Offset	0.0005512	0.00030525	1.8057		0.0798
Slope	0.000001235	1.2853×10^{-7}	9.6070		0.0000
		Variance Analysis			
Source	Sum of squares	D_f	Medium squares	F-statist.	Value p
Model	0.0000011472	1	0.0000011472	92.30	0.0000
The Rest	4.22607×10^{-7}	34	1.24296×10^{-8}		
Together (Corr.)	0.00000157	35			

The R^2 statistic shows that the model explains 73.1% of the variability in the energy consumption index of Conveyor A. The correlation coefficient (0.854862) indicates a moderately strong relationship between the variables. The residuals' standard deviation is 0.0001115. The mean absolute error (MAE) is 0.000079608, and the Durbin–Watson statistic suggests some serial correlation due to potential seasonal variations in the data.

For Conveyor B, multiple models were tested (logistic S curve, double hyperbolic, hyperbolic for the logarithm), but differences in correlation coefficients and R^2 values were minimal. The best correlation coefficient was 0.8616, while the hyperbolic model's correlation was slightly lower at 0.8603, making it the most comparable model, (Figure 9). The Equation (2) is as follows:

$$ZE_B = 1/(0.0004123 + 0.00000126\, Q_B) \tag{2}$$

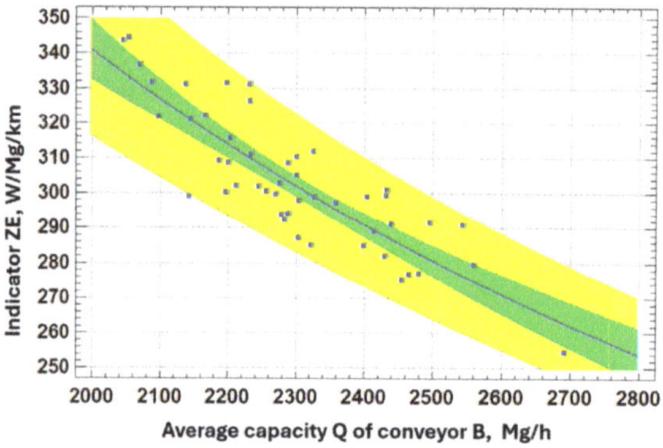

Figure 9. A simple regression of ZE for Conveyor B against average capacity Q.

The estimation analysis, including ANOVA, yielded the following:

- Correlation coefficient: 0.86031
- R-squared: 74.012%
- Standard error of estimation: 0.00010734
- Mean absolute error: 0.000089
- Durbin–Watson statistic: 0.58731 ($p = 0.0000$)
- Residual autocorrelation lag 1: 0.66312

Since the p-value in ANOVA (Table 6) is below 0.05, a statistically significant relationship exists between indicator ZE_B and Q_B at a 95% confidence level.

Table 6. Coefficients and variance analysis for selected regression model for Conveyor B (D_f—degrees of freedom).

	Least Squares Method	Standard	Statistics		Value
Parameter	Estimate	Error	T		p
Offset	0.00041227	0.00025337	1.6277		0.1105
Slope	0.0000012602	1.10102×10^{-7}	11.446		0.0000
		Variance Analysis			
Source	Sum of squares	D_f	Medium squares	F-statist.	Value p
Model	0.0000015094	1	0.0000015094	131.01	0.0000
The rest	5.29975×10^{-7}	46	1.15212×10^{-8}		
Together (corr.)	0.000002039	47			

When analyzing data from both conveyors together, the hyperbolic model ranked third after logarithmic and square root hyperbolic models. The minor differences in correlation coefficient (0.0001 lower) and R^2 index (0.07 lower) made the hyperbolic model suitable for comparative purposes. The Equation (3) is as follows:

$$ZE_{A\ and\ B} = 1/(0.00031769 + 0.0000013153\ Q_{A\ and\ B}) \quad (3)$$

The combined regression analysis resulted in the following:

- Correlation coefficient: 0.86312
- R-squared: 74.498%
- Standard error of estimation: 0.00011448
- Mean absolute error: 0.000088274
- Durbin–Watson statistic: 1.0102 ($p = 0.0000$)
- Residual autocorrelation Lag 1: 0.46853

The model explains 74.498% of the unit energy consumption indicators variability for both conveyors.

The selected regression curves for Conveyors A and B, and combined data were plotted on a standard graph (Figure 10). The regression curves run almost parallel, with minor differences in energy consumption indices. The difference for 2000 Mg/h is 9.9 Wh/Mg/km, reducing to 4.3 Wh/Mg/km at 2700 Mg/h, a decrease from 3% to 1.7%.

The hyperbolic model fits well, with point dispersion (forecast vs. observed) being mostly uniform, except for a few outliers (Figure 11). However, a variance of up to 30 Wh/Mg/km suggests that factors other than capacity influence energy consumption.

The scatter of results around the diagonal in the graph of observed and predicted values (Figure 12) and especially seasonal temperature variations appear to contribute significantly, with negative residuals in winter–spring and positive ones in summer–autumn (Figure 13). Future analysis will explore the influence of atmospheric conditions.

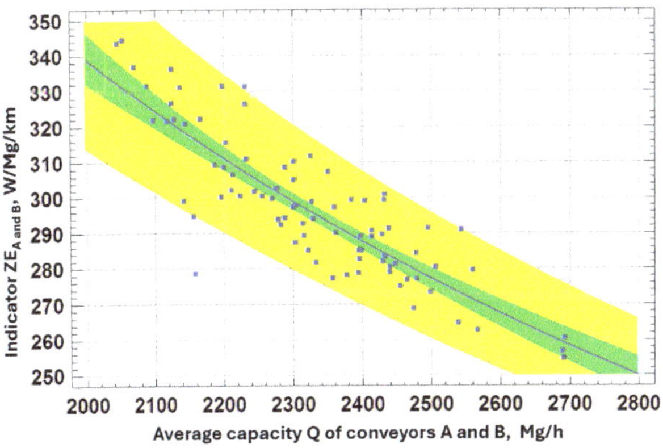

Figure 10. A simple regression of ZE for Conveyors A and B against average capacity Q.

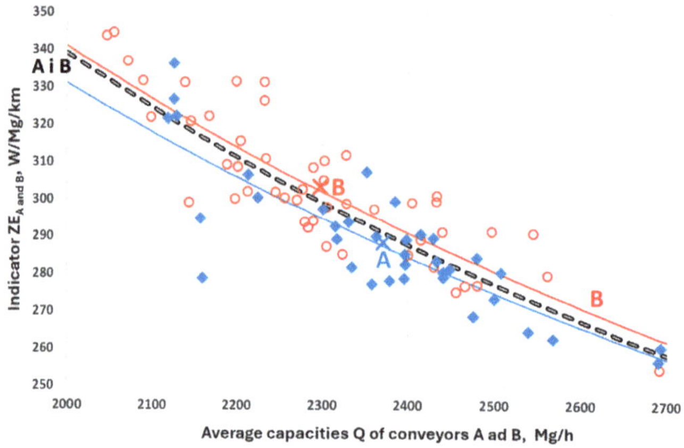

Figure 11. Comparison of regression models for Conveyors A (blue line), B (red line), and combined data (dashed line).

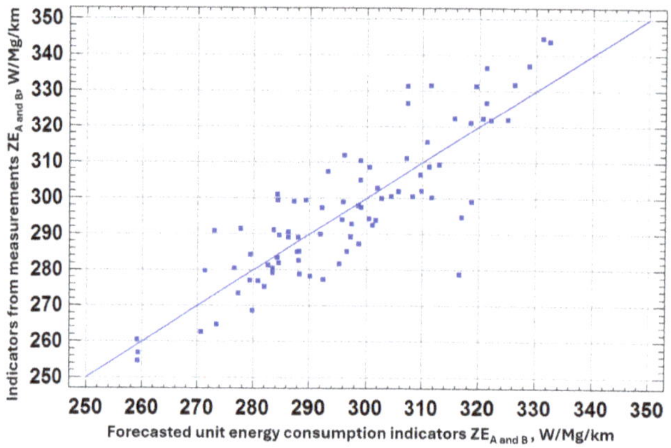

Figure 12. Comparison of predicted and observed values.

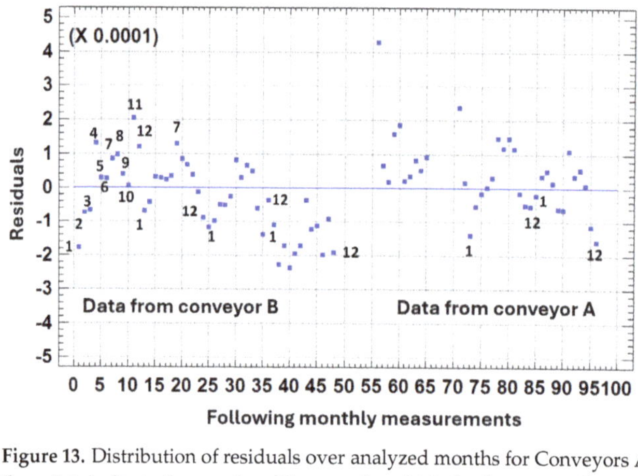

Figure 13. Distribution of residuals over analyzed months for Conveyors A and B. The numbers at the points indicate the months of the year for which the data was collected.

4. Conclusions

This study demonstrates that variations in conveyor energy consumption primarily result from differences in load distribution. A statistically significant disparity was observed between the average capacities of Conveyors A and B, with Conveyor A operating at 381.99 Mg/h and Conveyor B at 2309.47 Mg/h—an absolute difference of 72.51 Mg/h (3.04%).

A strong negative correlation was identified between ZE and conveyor capacity ($R^2 = -0.84$ and -0.85 for Conveyors A and B, respectively). A 35% relative increase in capacity (from 2000 up to 2700 Mg/h) led to a 26% reduction in ZE (from 340 down to 250 Wh/Mg/km), equating to a 90 Wh/Mg/km decrease. Proper load distribution could potentially reduce unit energy consumption by up to 30%.

Optimizing conveyor utilization—such as alternating higher load levels between conveyor lines—can yield substantial energy savings. For instance, instead of running both lines at 30% capacity daily, increasing one line's load to 60% while idling the other could further reduce energy consumption by as much as 150 Wh/Mg/km. These optimizations require no additional infrastructure investment, making them a cost-effective strategy for improving efficiency.

The Bełchatów power plant receives 30–40 million Mg of coal annually, transported over conveyor lines spanning several dozen kilometers. For 10 million Mg/year over 10 km, this equates to potential energy savings of 9 GWh (or 15 GWh at 60%). Scaling up to actual conditions—four times the coal volume and double the conveyor length—yields estimated savings of 72 GWh (or 120 GWh). Including overburden removal, these estimates could increase further (by multiplication by overburden to coal ratio + 1), emphasizing the potential of improved load management through centralized control in surface mines.

An additional opportunity lies in synchronizing conveyor operations with the daily solar cycle for overburden removal and coal extraction. Excess renewable energy, particularly from PV sources, often results in low or negative electricity prices and RES curtailments. Dynamic load control allows conveyor utilization to maximize this opportunity during peak solar production. A recent study in *Energies* explored this approach, demonstrating its feasibility in developing the Złoczew lignite deposit [14].

Future research will investigate additional energy-saving measures, such as energy-saving conveyor belts, and evaluate the benefits of load optimization and technological improvements. Further studies will also assess the impact of environmental factors, particularly temperature, on energy consumption variability, as preliminary findings suggest that up to 75% of ZE fluctuations stem from load variations.

Identifying the primary drivers of conveyor energy consumption under long-term industrial conditions enhances predictive accuracy and highlights significant opportunities for efficiency improvements. These findings provide practical guidelines for reducing operational costs and improving the sustainability of belt conveyor systems in lignite mining operations.

Author Contributions: Conceptualization, L.J. and M.B.; methodology, L.J.; software, L.J.; validation, M.B., L.J. and Z.K.; formal analysis, M.B.; investigation, Z.K.; resources, L.J. and M.B.; data curation, Z.K.; writing—original draft preparation, L.J.; writing—review and editing, M.B.; visualization, L.J.; supervision, M.B.; project administration, M.B.; funding acquisition, M.B. and L.J. All authors have read and agreed to the published version of the manuscript.

Funding: This research received no external funding.

Data Availability Statement: The data presented in this study are available on request from the corresponding author.

Acknowledgments: The authors would like to thank the mine management, for providing detailed data on the energy consumption of the analyzed conveyors.

Conflicts of Interest: Author Zbigniew Konieczka was employed by the PGE Polska Grupa Energetyczna S.A. The remaining authors declare that the research was conducted in the absence of any commercial or financial relationships that could be construed as a potential conflict of interest.

References

1. Soofastaei, A.; Karimpour, E.; Knights, P.; Kizil, M. Energy-efficient loading and hauling operations. In *Energy Efficiency in the Minerals Industry. Best Practices and Research Directions*; Springer: Berlin/Heidelberg, Germany, 2018; pp. 121–146. [CrossRef]
2. De Almeida, C.M.; de Castro Neves, T.; Arroyo, C.; Campos, P. Truck-and-loader versus conveyor belt system: An environmental and economic comparison. In Proceedings of the 27th International Symposium on Mine Planning and Equipment Selection-MPES 2018; Springer International Publishing: Cham, Switzerland, 2018; pp. 307–318.
3. Carter, R.A. Designing Today's Conveyor Systems for Tomorrow's Needs. *Eng. Min. J.* **2023**, *224*, 24–30.
4. Leonida, C. Moving Material More Efficiently. *Eng. Min. J.* **2023**, *224*, 32–37.
5. Bodziony, P.; Patyk, M. The influence of the mining operation environment on the energy consumption and technical availability of truck haulage operations in surface mines. *Energies* **2024**, *17*, 2654. [CrossRef]
6. Krysa, Z.; Bodziony, P.; Patyk, M. Exploitation of Mineral Resources in Conditions of Volatile Energy Prices: Technical and Economic Analysis of Low-Quality Deposits. *Energies* **2024**, *17*, 3379. [CrossRef]
7. Rakhmangulov, A.; Burmistrov, K.; Osintsev, N. Multi-criteria system's design methodology for selecting open pits dump trucks. *Sustainability* **2024**, *16*, 863. [CrossRef]
8. He, D.; Pang, Y.; Lodewijks, G. Green operations of belt conveyors by means of speed control. *Appl. Energy* **2017**, *188*, 330–341. [CrossRef]
9. Bajda, M.; Hardygóra, M.; Marasová, D. Energy efficiency of conveyor belts in raw materials industry. *Energies* **2022**, *15*, 3080. [CrossRef]
10. Szczeszek, P.; Jastrząb, A.; Karkoszka, T.; Prusko, A. Energy and ecological optimization of the belt conveyors' operation. *J. Achiev. Mater. Manuf. Eng.* **2023**, *121*, 18.
11. Mathaba, T.; Xia, X. Optimal and energy efficient operation of conveyor belt systems with downhill conveyors. *Energy Effic.* **2017**, *10*, 405–417. [CrossRef]
12. Kawalec, W.; Król, R.; Suchorab, N. Regenerative belt conveyor versus haul truck-based transport: Polish open-pit mines facing sustainable development challenges. *Sustainability* **2020**, *12*, 9215. [CrossRef]
13. Naworyta, W.; Sikora, M. Czy możliwa jest budowa elektrowni szczytowo-pompowej na terenach po wydobyciu węgla brunatnego? Analiza zagrożeń, ograniczeń i korzyści. Analiza zagrożeń, ograniczeń i korzyści. *Nowa Energ.* **2023**, *3*, 38–43. (In Polish)
14. Jurdziak, L.; Kawalec, W.; Kasztelewicz, Z.; Parczyk, P. Using RES Surpluses to Remove Overburden from Lignite Mines Can Improve the Nation's Energy Security. *Energies* **2024**, *18*, 104. [CrossRef]
15. Raj, P.P.; Pochont, N.R.; Teja, K.R.; Kumar, G.D. Solar PV Application in Industrial Conveyor System. *IOP Conf. Ser. Earth Environ. Sci.* **2020**, *573*, 012046.
16. Bajda, M.; Jurdziak, L.P. Electric Energy Needs of Belt Conveyors in One of the Polish Lignite Mines—Method of Analysis. *Civ. Environ. Eng. Rep.* **2025**, *35*, 170–184. [CrossRef]
17. Suchorab, N. Specific energy consumption–the comparison of belt conveyors. *Min. Sci.* **2019**, *26*, 263–274. [CrossRef]
18. Kawalec, W.; Suchorab, N.; Konieczna-Fuławka, M.; Król, R. Specific energy consumption of a belt conveyor system in a continuous surface mine. *Energies* **2020**, *13*, 5214. [CrossRef]
19. Bajda, M.; Jurdziak, L.; Pactwa, K.; Woźniak, J. Energy-Saving of Conveyor Belts in the Strategy and Reporting of Corporate Social Responsibility Initiatives of Producers. In *Global Congress on Manufacturing and Management*; Springer International Publishing: Cham, Switzerland, 2021; pp. 402–414.
20. Köken, E.; Lawal, A.I.; Onifade, M.; Özarslan, A. A comparative study on power calculation methods for conveyor belts in mining industry. *Int. J. Min. Reclam. Environ.* **2022**, *36*, 26–45. [CrossRef]
21. Suchorab-Matuszewska, N. Data-Driven Research on Belt Conveyors Energy Efficiency Classification. *Arch. Min. Sci.* **2024**, *69*, 375–389. [CrossRef]
22. Blazej, R.; Jurdziak, L.; Kawalec, W. *Energy Saving Solutions for Belt Conveyors in Lignite Surface Mines*; Institute of Mining Engineering, Wroclaw University of Science and Technology: Wrocław, Poland, 2015.
23. Roumpos, C.; Partsinevelos, P.; Agioutantis, Z.; Makantasis, K.; Vlachou, A. The optimal location of the distribution point of the belt conveyor system in continuous surface mining operations. *Simul. Model. Pract. Theory* **2014**, *47*, 19–27. [CrossRef]

24. Youssef, G. *Development of Conveyor Belts Design for Reducing Energy Consumption in Mining Applications*; Ain Shams University: Cairo, Egypt, 2016.
25. Zhou, S.J.; Zeng, F.; Du, J.; Wu, Q.; Ren, F. Experimental research on the energy consumption laws and its influencing factors of belt conveyor systems. In Proceedings of the 2016 13th International Computer Conference on Wavelet Active Media Technology and Information Processing (ICCWAMTIP), Chengdu, China, 16–18 December 2016; pp. 395–402.
26. Krol, R.; Kawalec, W.; Gladysiewicz, L. An effective belt conveyor for underground ore transportation systems. *IOP Conf. Ser. Earth Environ. Sci.* **2017**, *95*, 042047. [CrossRef]
27. Munzenberger, P.; Wheeler, C. Laboratory measurement of the indentation rolling resistance of conveyor belts. *Measurement* **2016**, *94*, 909–918. [CrossRef]
28. Woźniak, D. Laboratory tests of indentation rolling resistance of conveyor belts. *Measurement* **2020**, *150*, 107065. [CrossRef]
29. dos Santos e Santos, L.; Ribeiro Filho, P.R.C.F.; Macêdo, E.N. Indentation rolling resistance in pipe conveyor belts: A review. *J. Braz. Soc. Mech. Sci. Eng.* **2021**, *43*, 230. [CrossRef]
30. Zhou, L.; Wu, Z.; Li, Y.; Yao, H.; Liu, Y.; Yuan, Y.; Cao, X. Temperature characteristics of indentation rolling resistance of belt conveyor. *J. Mech. Sci. Technol.* **2023**, *37*, 4125–4135. [CrossRef]
31. Wan, L.; Lin, F. Simulation and Experimental Study on Indentation Rolling Resistance of a Belt Conveyor. *Machines* **2024**, *12*, 778. [CrossRef]
32. Kulinowski, P.; Kasza, P.; Zarzycki, J. Influence of design parameters of idler bearing units on the energy consumption of a belt conveyor. *Sustainability* **2021**, *13*, 437. [CrossRef]
33. Kulinowski, P.; Kasza, P.; Zarzycki, J. Methods of Testing of Roller Rotational Resistance in Aspect of Energy Consumption of a Belt Conveyor. *Energies* **2022**, *16*, 26. [CrossRef]
34. Malakhov, V.A.; Dyachenko, V. Rolling resistance coefficient of belt conveyor rollers as function of operating conditions in mines. *Eurasian Min.* **2022**, *1*, 67–71. [CrossRef]
35. Opasiak, T.; Margielewicz, J.; Gąska, D.; Haniszewski, T. Rollers for belt conveyors in terms of rotation resistance and energy efficiency. *Transp. Probl.* **2022**, *17*, 57–68. [CrossRef]
36. Bajda, M.; Hardygóra, M. Analysis of the Influence of the Type of Belt on the Energy Consumption of Transport Processes in a Belt Conveyor. *Energies* **2021**, *14*, 6180. [CrossRef]
37. Jena, M.C.; Mishra, S.K.; Moharana, H.S. Experimental investigation on maximizing conveying efficiency of belt conveyors used in series application and estimation of power consumption through statistical analysis. *Environ. Prog. Sustain. Energy* **2023**, *42*, e14031. [CrossRef]
38. Wen, L.; Liang, B.; Zhang, L.; Hao, B.; Yang, Z. Research on coal volume detection and energy-saving optimization intelligent control method of belt conveyor based on laser and binocular visual fusion. *IEEE Access* **2023**, *12*, 75238–75248. [CrossRef]
39. Dinolov, O.; Yordanov, R. Tailored Review on the Research in the Electric-Energy Efficiency of Transport Processes (Part 1): Bulk Materials and Fluids. In Proceedings of the 2024 9th International Conference on Energy Efficiency and Agricultural Engineering (EE&AE), Ruse, Bulgaria, 27–29 June 2024; pp. 1–9.
40. Lodewijks, G.; Schott, D.L.; Pang, Y. Energy saving at belt conveyors by speed control. In Proceedings of the Beltcon 16 Conference, Johannesburg, South Africa, 3–4 August 2011; pp. 1–10.
41. Ji, J.; Miao, C.; Li, X.; Liu, Y. Speed regulation strategy and algorithm for the variable-belt-speed energy-saving control of a belt conveyor based on the material flow rate. *PLoS ONE* **2021**, *16*, e0247279. [CrossRef] [PubMed]
42. Ni, Y.; Chen, L. Research of energy saving control System of Mine belt conveyors. In Proceedings of the 2021 International Conference on Intelligent Transportation, Big Data & Smart City (ICITBS), Xi'an, China, 27–28 March 2021; IEEE Access: Piscataway, NJ, USA, 2017; pp. 579–581.
43. Ilic, D.; Wheeler, C. Measurement and simulation of the bulk solid load on a conveyor belt during transportation. *Powder Technol.* **2017**, *307*, 190–202. [CrossRef]
44. Wheeler, C.; Munzenberger, P.; Ausling, D.; Beh, B. *Energy Efficient Belt Conveyor Design*; Beltcon 18-11, IMHC Beltcon, South Africa, 2018; 27p. Available online: https://www.beltcon.org.za/archives/ (accessed on 9 April 2025).
45. Doroszuk, B.; Krol, R. Analysis of conveyor belt wear caused by material acceleration in transfer stations. *Min. Sci.* **2019**, *26*, 189–201.
46. Doroszuk, B.; Król, R.; Wajs, J. Simple design solution for harsh operating conditions: Redesign of conveyor transfer station with reverse engineering and DEM simulations. *Energies* **2021**, *14*, 4008. [CrossRef]
47. Ambriško, L'.; Marasová, D., Jr.; Klapko, P. Energy Balance of the Dynamic Impact Stressing of Conveyor Belts. *Appl. Sci.* **2023**, *13*, 4104. [CrossRef]

energies

MDPI

Article

Copper as a Critical Resource in the Energy Transition

Marta Podobińska-Staniec *, Anna Wiktor-Sułkowska, Arkadiusz Kustra and Sylwia Lorenc-Szot

Economics and Management of Raw Materials and Energy Transition Group, Faculty of Civil Engineering and Resource Management, AGH University, 30-059 Krakow, Poland; wiktor@agh.edu.pl (A.W.-S.); kustra@agh.edu.pl (A.K.); slorenc@agh.edu.pl (S.L.-S.)

* Correspondence: mstaniec@agh.edu.pl

Abstract: The energy transition requires significant amounts of critical raw materials, with demand projected to continue increasing. Analyses indicate that the supply of these materials will fall short of the requirements for the energy transition plans based on renewable energy sources. To address this challenge, it is essential to examine the global value chains of critical raw materials and their intermediates, identifying the risks linked to concentrating downstream processes in countries that may not operate under market-oriented principles. Such countries could become "price makers" on international markets, influencing costs and availability. This article focuses on copper, a key raw material for the energy transition, and explores its global value chains in detail. It highlights the risks associated with copper extraction in regions with the largest identified deposits, such as South America and Africa, as well as challenges related to smelting and refining, which are predominantly carried out in China. Additionally, the article presents an analysis of the operations of KGHM Polska Miedź S.A., one of Europe's largest copper producers, headquartered in Poland but also active in North and South America.

Keywords: energy transition; critical raw materials; global value chains; copper products

1. Introduction

Faced with the challenges of climate change, in the past decade, decarbonization and the energy transition have become key priorities and major challenges for many countries and communities around the world. One of the key aspects of the energy transition, for which the main projected values have been calculated for the period up to 2050, is the replacement of generation of energy from fossil fuels, including oil, gas and coal, with generation from renewable resources, such as wind, hydroelectric and solar power. In the EU, this development strategy to transform the European Union into a climate-neutral area has, since December 2019, been referred to as the European Green Deal. The strategy has more ambitious goals for 2030 and 2050. It calls for a 55% reduction in greenhouse gas emissions by 2023 compared to 1990, zero greenhouse gas emissions by 2050, and a shift away from fossil fuels to implementation of sustainable and clean energy with prioritization of the area of energy efficiency improvements [1]. The indicated goals seem extremely important, especially for hard to decarbonize industries where it is particularly difficult or impossible to move away from coal in favour of renewable energy, e.g., metallurgy, transportation, chemicals, etc.

As Xu et al. point out, the energy transition refers both to the reconstruction of the energy system from energy production to energy use, but also includes the transformation of economic and social systems brought about by the digitization and intelligence of new scenarios for energy management, transportation and use [2]. It is pointed out that the motivation for the energy transition is not only to reduce climate change and the environmental impact of the energy sector, but also to improve energy efficiency, ensure energy security, maintain the reliability of energy access, reduce costs, and promote sustainable development [3,4]. However, achieving these ambitious goals requires not only advanced technologies, but also access to key, sustainable raw materials needed for clean energy generation technologies, as well as digital, space or defence technologies [5]. In advanced technologies, a key role is played by devices that use elements containing elements classified as CRMs (critical raw materials)—the raw materials whose economic importance to the EU is growing year by year, and which are inherently exhaustible. These raw materials have a strategic impact on the energy transition (they are used in photovoltaic cells, electrolysers, synthetic fuels, wind turbines, and batteries or fuel cells, among others), but due to the scarcity of geological resources, their concentration in different regions of the world, geopolitical issues, trade policy issues or the lack of good, affordable substitutes, their supply is very likely to be disrupted. In this context, consideration of critical raw materials (CRMs) and their relationship to the risk of supply chain disruption, appears to be one of the key issues for achieving a global green energy transition. According to the World Bank, the list of critical minerals that are essential for the transition to green energy mainly includes aluminium, chromium, cobalt, copper, graphite, indium, iron, lead, lithium, manganese, molybdenum, neodymium, nickel, silver, titanium, vanadium and zinc [6].

Of particular importance in this set is copper, which, due to its properties, is fundamental for the energy transition. All the technologies considered in this context, whether related to photovoltaics, wind farms, energy storage or electromobility, consume copper. According to a Goldman Sachs report [7], the critical role that copper plays in achieving climate goals cannot be overstated. Due to its electrical conductivity and low reactivity, copper is the most cost-effective material for renewable energy storage and transportation. Based on the Goldman Sachs report [7], copper is the 'new oil' of the 21st century, and complete decarbonization will not be possible without this commodity. According to forecasts, the green transition will contribute to a surge in demand for copper: analysts estimate that the copper demand associated with the energy transition will increase by nearly 600% by 2030 to 5.4 million tons in the baseline scenario (with rapid and intensive deployment of green technologies, the demand could increase by as much as 900% to 8.7 million tons) [7].

This article aims to contribute to the discussion on the energy transition currently taking place in the world, especially in Europe. The role of critical raw materials in the process of change is highlighted, and the risks that are associated with the acquisition of raw materials and the possible lack thereof for the development of renewable technologies for generating electricity are noted.

Special attention was paid to the role of copper in the energy transition and its acquisition, along with identification of the copper value chain worldwide.

Since Poland also has copper ore reserves and is a significant producer, the country's copper value chain was also identified and the exports and imports of products that arise at the main stages of this chain were analysed. In order to practically illustrate the phenomena occurring in the industry, the Polish copper producer KGHM Polska Miedź SA was analysed as a provider of strategic raw materials and other materials for low carbon and renewable installations providing capabilities for the transformation and energy security at the same time.

The research questions that the authors of this article seek to answer are as follows:

- What is the role of copper used in technologies for the energy transition, and how can the growing demand be met while addressing the supply gap?
- How are the copper products that form the particular value chain stages traded on world markets?
- How does Poland as a country provide copper products for the energy transition and how is this activity organized through KGHM SA?

2. Materials and Methods

To determine the current state of the art, a literature search was conducted by looking for papers related to the issue of "Energy transformation" in the Scopus database. This search showed that, over the years, the term has appeared as many as 4823 times, mostly since 2020. To examine the literature of the subject matter covered in this paper, the Scopus database was searched (with an initial result of 4823 papers) according to the Prisma method presented in Figure 1. This process finally yielded 45 papers published since 2020 and indicated the existence of a secondary source of literature, which is documents issued by European decision-making units.

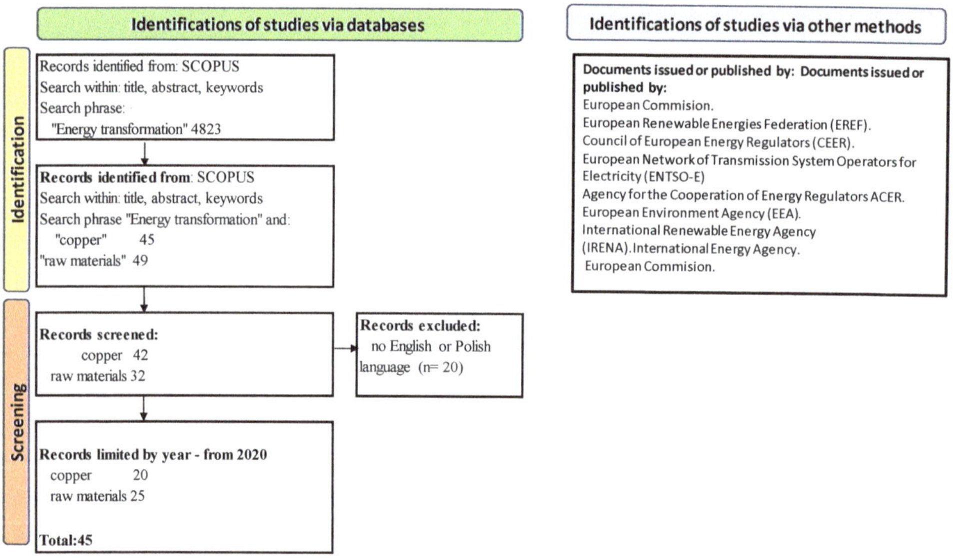

Figure 1. Literature review—outline.

The growing global demand for energy and environmental restrictions on the use of fossil fuels have prompted intensive research into sustainable, clean, low-cost and, above all, environmentally friendly energy sources. Most of the selected papers deal with topics related to the study of the chemical properties of minerals for the purpose of better generation or conduction of new generation energy, including [8–17].

Climate change and the energy transition have caused massive changes in the energy system, leading to a great degree of uncertainty around materials, energy and carbon emissions in the power generation infrastructure. Chinese researchers are well aware of this fact, with numerous studies analysing the links between materials, energy and carbon emissions in China [18–25].

The effective use of critical raw materials while ensuring optimal economic conditions in countries is examined in two studies [26,27].

It is also possible to distinguish a group of studies in which the use of copper in the context of energy creation and storage is central. These include work investigating tetrapyrrole ligands, such as porphyrin-, phthalocyanine- and corrole-based systems, which are versatile and can catalyse ORR very efficiently when bound to metals such as Fe, Co and Cu [28]. Going further, in [29,30], researchers studied copper nanoparticles, which are highly valued in engineering and technological research due to their wide potential applications. In addition, the study [31] includes an examination of insulating materials, for example, when using copper to manufacture the aforementioned electric motors.

This article continues with an analysis of the energy transformation and demand for copper due to this process. A detailed analysis of the main global copper value chains was carried out, taking into account copper's production, processing and distribution. Next, the role of KGHM Polska Miedź S.A., one of the largest copper companies in the world, as a key link in the global copper supply chain, including supplying high-purity copper for advanced energy technologies, was analysed.

3. Results

3.1. Demand for Copper Due to the Energy Transition

The Energy Transition Commission [32] estimates that global electricity consumption in 2050 will quadruple in comparison with the consumption in 2022, i.e., from 28,000 TWh to 110,000 TWh.

The assumptions are that 75 percent of energy in 2050 will be supplied by wind and solar power, with the rest of the demand met by nuclear, hydroelectric and other zero-carbon sources.

Figure below (Figure 2) shows the necessary changes that must occur by 2050 to meet the energy transition goals. These developments are discussed below in the context of the demand for copper:

Energy technologies	Wind	Solar	Power grid	Green Hydrogen	EV's and baterries
2022	1 TW	1.2 TW	70 Million Km	<1 MT	25 M EV FLEET
2050	14-15 TW	26-34 TW	200 Million Km	500-800 MT	1500 M
Increse 2022-2050	x15	x25	x3	x500	x60

Figure 2. Changes in energy mix in 2050 compared to 2022.

- The current installed capacity for wind power is 1 TW, and it will have to increase about fifteen times to 14–15 TW. An average of 4.7 t of Cu is used for building a single 3 MW wind turbine. Copper is also needed for offshore wind power, as its specific gravity, compared to aluminum, makes it easier to install underwater connections.
- The installed capacity for solar power is 1.2 TW and will have to increase about twenty-five times to 26–34 TW. An average of 3.8 t of Cu must be used for building a 1 MW solar power plant.
- The power grid will have to be expanded from the current 75 million km to 200 million km. Copper would be best used for expanding the power grid, as it is one of the

metals with the best electrical conductivity, which offsets losses in the transmission itself, and its thermal conductivity is more than half that of aluminum.

- Low-carbon hydrogen consumption will have to increase from the current level of 1 MT to as much as 500–800 MT.
- By 2050, the plan is to almost completely replace internal combustion cars with electric ones, so more than 1.5 billion EVs should be produced, as well as about 200 million electric trucks and buses. This requires a total battery capacity of up to 150 TWh. For EVs, copper is needed for batteries, electrical components, and in charging stations. Electric cars contain an average of two to three times more copper than their counterparts with internal combustion engines Figure 3. It is also difficult here to talk about substituting copper with, for example, aluminum. Copper does not heat up as aluminum does, and the reductions in energy consumption when using copper reach about 25%.

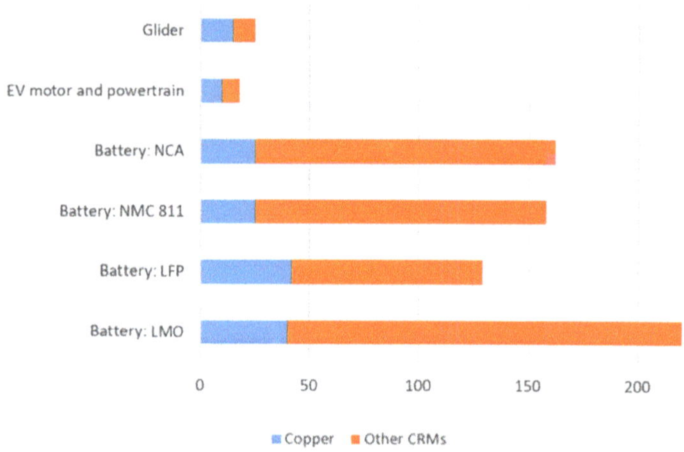

Figure 3. Concentration of critical materials used in battery vehicles [kg/vehicle].

Figure 4 show the consumption of critical raw materials in the technologies that are to contribute to the energy transition. In the chart, it can be seen that each technology entails significant copper consumption. To build an offshore wind power plant, about eight times as many tons of copper per 1 MW are needed in comparison with a conventional power plant. Additions of onshore wind farms and solar power plants require two to three times more copper per 1 MW of installed capacity compared to conventional power plants [33] (Figure 5).

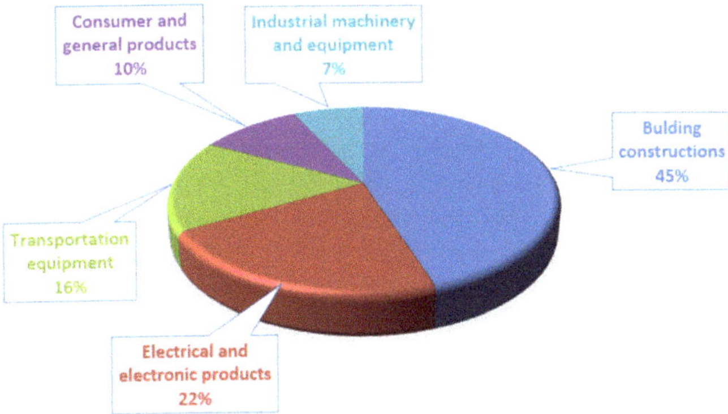

Figure 4. Total copper demand structure in 2023.

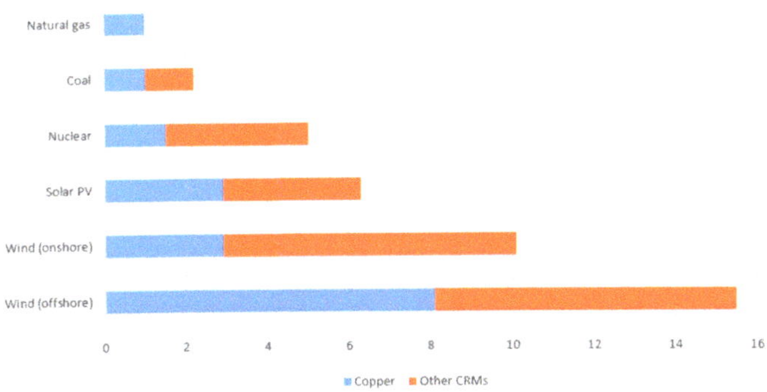

Figure 5. Consumption of critical raw materials in the technologies that are to contribute to the energy transition [tons/MW].

Figure 3 shows a summary of the consumption of critical raw materials used for building various batteries produced for EVs and compares the consumption of these raw materials for the motor and power trains, as well as the glider in EVs. This chart shows that, on average, to build an EV requires ca. 36 kg of copper per battery, ca. 16 kg per EV motor and powertrain and ca. 18 kg per glider, which makes ca. 70 kg of copper in total. For internal combustion engines (ICE) cars, the copper consumption is ca. 15 kg for the internal combustion engine and powertrain and ca. 20 kg for the glider, which makes 35 kg in total. That is, it takes at least twice as much copper to produce one EV.

Due to the energy transition, as much as 22 million tons of copper should be supplied to the market by 2050, a figure equal to the entire world's output in 2023. At this point, it should also be emphasized that 22 million tons of copper by 2050 is only the copper needed to implement the energy transition targets. The Energy Transition Commission (ETC) estimates that annual output should rise to a maximum of 34 million t by 2030 under basic decarbonization assumptions, and a minimum of 30 million t with additional increases in efficiency and recycling rates [32]. In addition, copper is used in other sectors, i.e., building construction, transportation equipment, consumer and general products, and industrial machinery (Figure 4).

ETC estimates that, to meet the needs of all sectors, the cumulative demand for copper by 2050 should be 1135 million t, with estimated available reserves of 1000 million t and resources of 5600 million t. This means that at present (2024), there is still a copper shortfall of 135 million tons.

3.2. Filling the Copper Supply Gap Through Investments

As a result, it is necessary to increase reserves. Such increases take place all the time. Figure 6 below shows how reserves have changed over the past ten years (2013–2023) from 600 million tons to 1000 million tons. Also shown is the change from 2022 to 2023, when reserves increased by 110 million tons. This value shows that, in the near future, reserves may change so that the currently expected supply gap will perhaps be completely closed in a year's time. The chart below shows that the largest reserves are located in Chile, Peru, Australia and Russia [34–36]. Poland ranks eighth in terms of domestic resources (34 million tons). KGHM, in turn, has 40 million t of copper reserves at all of its mines [37].

Figure 6. Global copper reserves [thousand metric tons].

The data on reserves (Figure 6) show where they can be expanded and where production can be increased. In Europe, the largest and most significant supplier of copper is KGHM (output 445.5 thousand tons, refining 592 thousand tons). The chart shows that Polish reserves are growing, with the latest value of 34 million tons. In 2023, domestic reserves grew by 13%. KGHM, in its 2022 report [34], states that it controls 40 million t of resources within the group.

With the huge demand for copper due to the energy transition, the reserves, as well as the mining and production capacity on the Old Continent, must be steadily increased. This will reduce imports and, in turn, significantly reduce global CO_2 emissions. The carbon footprint of domestic production is lower than in Asia. Local production will also reduce the carbon footprint associated with transportation. Compared to 1990 levels, Europe's copper industry has reduced unit energy consumption by 60%, and emissions from the production of the metal in Europe now amount to just 0.4% of the EU's total greenhouse gas emissions [38].

To be able to increase reserves, mining capacity and, as a result, Cu production capital expenditures must be continuously incurred and increased. Figure 7 below shows the spending pattern for selected critical raw materials related to the energy transition from 2008 to 2022.

Figure 7. Exploration budget for selected critical minerals [USD million].

Figure 7 demonstrates that, globally, the largest critical capital expenditures on raw materials (CRMs) are incurred for copper exploration. Since 2016, there has been a gradual increase in spending. There was a decline after 2019, which is likely related to the COVID-19 pandemic. In 2022, the increase in investment in copper exploration was 21%, in comparison with the previous year, and the level of spending was the highest since 2014 at about 3000 USD million [33].

Figure 8 shows the structure of capital expenditure on copper exploration by country. In 2022, Latin America (Chile, Peru) had the largest copper exploration expenditures, followed closely by Australia and Canada. The chart shows that, compared to 2021, all of the mentioned countries increased their spending in 2022—Latin America and Australia by more than 30%, and the US and Canada by approximately 15%. In Africa, there was a slight increase, mainly due to the DRC and Ecuador [33]. However, the average annual increase in capital expenditure on copper exploration does not translate into discoveries of new deposits, the number of which has been declining steadily since 2007. Figure 9 shows the number of newly discovered deposits in the context of copper prices on world markets.

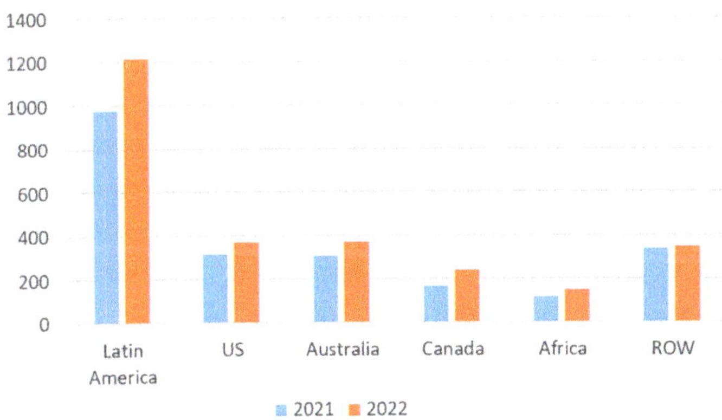

Figure 8. Copper exploration budget by country [USD million].

Figure 9. Major copper discoveries 2004–2023.

Exploration-related capital expenditures, shown in Figure 9, translate into global copper production, which are a consequence of exploration processes. In 2023, 22 million tons of copper was mined worldwide, 70% of which was mined by five countries: Chile (5 million tons), Peru (2.6 million tons), Congo (2.5 million tons), China (1.7 million), the US (1.1 million tons) [36] (Figure 10).

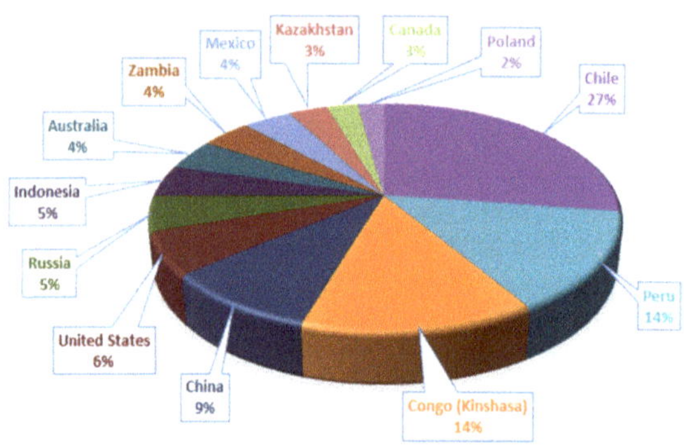

Figure 10. World copper mine production in 2023.

3.3. Filling the Copper Supply Gap by Recycling—Urban Mining

It is aimed to fill the copper supply gap by recycling. It is estimated that up to 80% of the copper produced to date is still in production use [38]. In theory, copper is a metal that can be recycled many times and still maintain its properties. Currently, 60% of end-of-life copper is recycled. The energy transition assumptions assume that the figure should be 90% by 2040.

It has already been mentioned in this article that, as part of the energy transition, by 2050, 22 million tons of copper will have been used mainly in wind turbines, solar panels, transmission grids, electric cars and all the infrastructure to support them. The current energy transition has been underway since 2019, and some copper has already entered the energy market. However, by 2030, small amounts of copper will have become recoverable from equipment used for the transition itself. The average lifespan of wind turbines and solar panels is 25 years, while electric cars have an average lifespan of 15 years. That is, the first electric appliances from the energy transition will be able to be processed after 2030, and most of them only after 2040. Therefore, it is necessary to look for copper-containing raw materials in other sources, i.e., in the already-built power system and outside it, i.e., in transportation, buildings, equipment, flotation tailings, etc.

Recycling has much potential to achieve a reduction in the demand for raw material extraction, but the whole process needs considerable improvement. Several years ago, it was estimated that up to 700 million unused devices were unclaimed across the EU, and among EVs that had reached their EOL (end of life), 3.5 million had not returned to the EU closed loop [33,39]. Recycling should be based on efficient collection of waste, so perhaps it would be worthwhile to introduce economic incentives that would make people take action themselves to return, for example, e-waste, to the right place. An infrastructure accessible to everyone should be built for the proper collection of such waste, where it would be subject to demolition and selection. The IEA estimates that by 2040, the recovered CRM volumes from used batteries could reduce the overall copper demand by up to 10% [40]. The energy transition's circular economy activities can achieve more than 7 MT of copper by 2050, which should meet more than 40% of the total energy transition demand [32].

4. Discussion

In light of the above-presented trends concerning copper demand driven by the energy transition, the next step involved analysing the main supply chains of this resource.

The tables (Tables 1–3) indicate the value migration associated with the trade in copper products, arising at different stages of the entire value chain. Considering the upstream processes and the ores and concentrates produced therein, it should be noted that the world's largest exporters were countries with the largest copper ore deposits. Thus, South America is dominant here, represented especially by Chile and Brazil. The shares of Asia, i.e., Indonesia and Kazakhstan, and Australia, are also significant. In turn, the largest buyers of copper ore and imported concentrate are countries that are more technologically advanced and that are engaged in further processing and production of subsequent products in the value chain. These include China, Japan and Germany, but also Scandinavian countries such as Sweden and Finland, as well as emerging Asian markets such as the Philippines and India.

Table 1. Exports and imports of copper ores and concentrates in 2023 [41].

Exports of Copper Ores and Concentrates			Imports of Copper Ores and Concentrates		
		Thousand USD			**Thousand USD**
1	Chile	24,322,170	1	China	59,941,382
2	Peru	19,991,449	2	Japan	11,740,694
3	Indonesia	8,326,476	3	Germany	3,306,776
4	Australia	4,344,555	4	Philippines	3,265,666
5	Mexico	3,715,931	5	India	3,057,890
6	Brazil	3,465,813	6	Spain	2,486,153
7	Canada	3,185,231	7	Bulgaria	1,938,247
8	Kazakhstan	3,068,322	8	Finland	916,922
9	United States	2,607,241	9	Sweden	913,885
10	Panama	2,469,049	10	Canada	719,882

Table 2. Exports and imports of unrefined copper and copper anodes for electrolysis in 2023 [41].

Exports of Unrefined Copper, Copper Anodes for Electrolysis			Import of Unrefined Copper, Copper Anodes for Electrolysis		
		Thousand USD			Thousand USD
1	Zambia	5,030,931	1	China	8,418,061
2	Chile	2,134,844	2	India	2,009,002
3	Sweden	1,325,665	3	Belgium	1,763,111
4	Bulgaria	962,632	4	Canada	1,276,751
5	South Africa	367,941	5	Finland	305,274
6	United States	244,559	6	Austria	216,312
7	Slovak Republic	215,121	7	Kazakhstan	109,694
8	Spain	207,820	8	Germany	66,051
9	Peru	207,674	9	Australia	56,131
10	Philippines	202,518	10	Serbia	48,566

Table 3. Exports and imports of copper cathodes and cathode components in 2023 [41].

Exports of Copper Cathodes and Cathode Components			Imports of Copper Cathodes and Cathode Components		
		Thousand US			Thousand USD
1	Chile	16,948,007	1	China	30,110,902
2	Japan	5,511,558	2	United States	6,671,446
3	Australia	3,370,027	3	Italy	4,888,128
4	Kazakhstan	2,840,465	4	Germany	3,922,802
5	Peru	2,689,386	5	Turkey	3,596,195
6	Poland	2,565,562	6	United Arab Emirates	3,386,355
7	China	2,421,337	7	Thailand	2,988,155
8	Philippines	1,941,133	8	India	2,793,250
9	Bulgaria	1,614,014	9	Malaysia	2,037,671
10	Zambia	1,552,586	10	Brazil	1,992,967

Unrefined copper and anodes for electrolysis are produced and exported by producers who have already sold ores and concentrates at an earlier stage (Chile, Peru, the United States). One can also distinguish countries that imported ores and concentrates and engaged in their further processing, such as Sweden, Bulgaria, Spain and the Philippines. However, the dominant country in the set of exporters is Zambia, which has significant copper ore reserves but is not a significant player at the stage of ore extraction and concentrate production.

When it comes to imports of unrefined copper and anodes, the highest value is again shown by China, but also by countries which previously imported fewer processed products in the form of ore or concentrates, including India, Germany, Finland, and Canada.

As for exports of more processed copper cathodes, the largest exporters are countries that have significant deposits and sell copper products at earlier stages of processing: Chile,

Peru, and Zambia. Exporters include countries that import fewer processed products, instead increasing the value of early-stage products and exporting them to other countries, thereby earning the margin generated from the increase in value. These include Japan, the Philippines, Bulgaria, Kazakhstan, and China. The list of exporters also includes Poland, which had not previously appeared in the lists, but is known to be a European player that produces ores and concentrates but does not sell significant quantities of them, and is trying to lengthen the value chain by selling and exporting more processed products, such as cathodes.

Importers are dominated by China again, but also includes countries such as Germany and India, which have already imported other products. There are also countries such as Italy, Turkey, Thailand, Malaysia which have the capacity to conduct further processing and produce alloys and other copper products for which cathodes are needed.

This analysis of the value of exports and imports in each country shows that China is the dominant country, which, having no ore deposits, buys the respective products created at subsequent stages of the value chain as well as processes them. In this way, profitability of the implemented processes is achieved by offering products that provide higher margins. This is illustrated by the flow chart for the copper value chain prepared by the Fraunhofer Institute, which features a strong increase in the value stream over the subsequent stages of the chain [42].

In turn, South American countries, especially Chile, Peru, and Brazil, but also Latin American counties such as Panama and Mexico, have a strong value flow in the upstream in the stages of ore extraction and concentrate production. As the value chain progresses, the stream becomes smaller and smaller. This is confirmed by a diagram prepared by the Fraunhofer Institute [42]. It confirms that more advanced production is performed by the countries with more innovative technologies and solutions, and at the same time these countries take over the margins in these value chains.

The instability of the geopolitical situation of the European Union countries and the disruption to the stability of the supply of energy resources have led to a reorientation of strategic measures in the area of energy security. For years, in legal terms, the European Parliament has been issuing directives on green governance and indicating emission standards. The RED III Directive (2023/2413) and the EED Energy Efficiency Directive (2023/1791), mentioned in the introduction to this paper, provide some of the basic standards. In 2022, the European Union introduced the RE-Power EU plan, the main objective of which is to diversify supply and accelerate the green transition process to enhance energy security [43].

To meet these challenges, the European Union continues to introduce new initiatives to help achieve the goals of the green transition and minimize the risk of disruptions in the supply of critical raw materials necessary for the transition. In the second half of 2023, two key pieces of legislation came into effect: the RED III Renewable Energy Directive (2023/2413) and the EED Energy Efficiency Directive (2023/1791) [43,44]. In the first half of 2024, while implementing these directives and regulations, the European Union has set specific targets for 2030, including increasing the share of renewable energy in total energy consumption to 42.5%, improving energy efficiency, achieving climate neutrality, and recycling, processing, or extracting strategic raw materials. From the point of view of the pursuit of a seamless transition, the European Parliament's Regulation, in particular, appears to be particularly important, as it represents a significant step towards ensuring the stability and sustainable supply of critical raw materials for the EU's economy. Implementation of the Act's objectives should strengthen the supply of these raw materials from the EU's own sources and reduce dependence on individual external suppliers. It is also intended to encourage the development of new strategic projects for the extraction, processing, recycling of and substitution for any of the strategic raw materials [45].

Copper Production in Poland vs. Global Production

Given the importance of copper in the energy transition, it is also a significant resource for Poland, especially since Poland is a major producer in Europe. In Poland, copper-bearing deposits are located in the western part of the country, and are mined by KGHM Polska Miedź S.A. (KGHM). Considering the copper value chain which contributes to the energy transition, KGHM will be among the companies operating seven of the eight stages of the chain, i.e., exploitation and assessment, ore extraction, ore enrichment, smelting, refining, processing and recycling (Figure 11). Therefore, it is worth analysing the situation of Poland in a global context and consider the requirements of the energy transition in order to be able to draw conclusions about the feasibility of projects aimed at reducing CO_2 emissions.

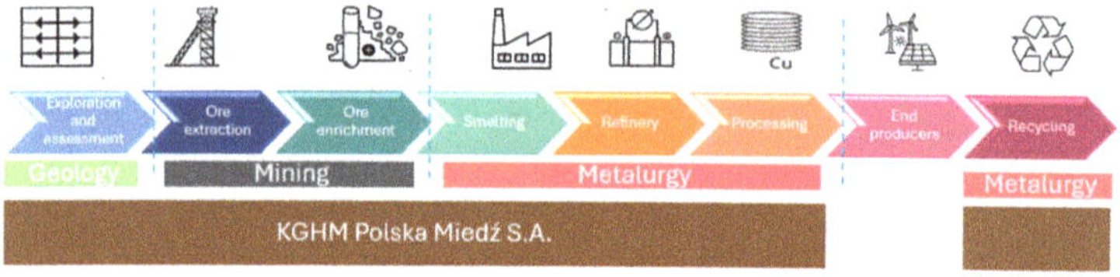

Figure 11. KGHM Group activities in the copper value chain for energy transition needs.

In Poland, KGHM mined 2% of the world's Cu output. Mining is carried out at the three Lubin mines, with a total of 72,300 tons mined in 2023, as well as at Polkowice-Sieroszowice (198.6 thousand tons), and Rudna mine (174.5 thousand tons). The aggregate output from the three sites adds up to 445,500 t, and this volume ranks Poland 13th in the United States Geological Survey (USGS) (Figure 6). In addition, KGHM also exploits foreign deposits. KGHM International LTD (a subsidiary of KGHM Polska Miedź S.A.) exploits copper bearing deposits in the United States; these are the Robinson mines (31.5 thousand t) and Carlota (3.9 thousand t). It has additional deposits in Canada, in the Sudbury Basin (4500 t) and holds a 55% stake in mining company Sierra Gorda S.C.M. in Chile (143,000 t) [37]. The sum of the Polish company's domestic and foreign output is 628.3 thousand t, which is nearly 3% of the world's output.

Copper ore mined in Poland contains an average of 1.5% Cu, so the next step is enrichment, or mechanical processing of the ore, which results in a concentrate with a higher Cu content. The concentrate is sent to the copper smelters HM Głogów and HM Legnica, where two other stages of smelting and refining follow. In smelters, in addition to their own concentrate, the input consists of purchased copper-bearing materials (concentrates, copper scrap, blister copper). In 2023, HM Głogów produced 470.1 thousand t of electrolytic (cathode) copper, and HM Legnica produced 122.3 thousand t, of which copper scrap accounted for 25% of electrolytic copper production (142.2 thousand t Cu). A total of 592,000 t of Cu were produced in Poland in 2023. Figure 12 below shows that, over the past four years (2019–2023), the upward trend in domestic copper production continued. The volume of ore remained at a similar level, but the copper content was higher. Foreign deposits have seen a decline over the past three years, mainly due to the sale of the Chilean Franke mine in 2022, lower production at the Robinson mine and lower copper content in ore at Sierra Gorda S.C.M. In 2023, the KGHM Group produced a total of 711,000 tons of Cu.

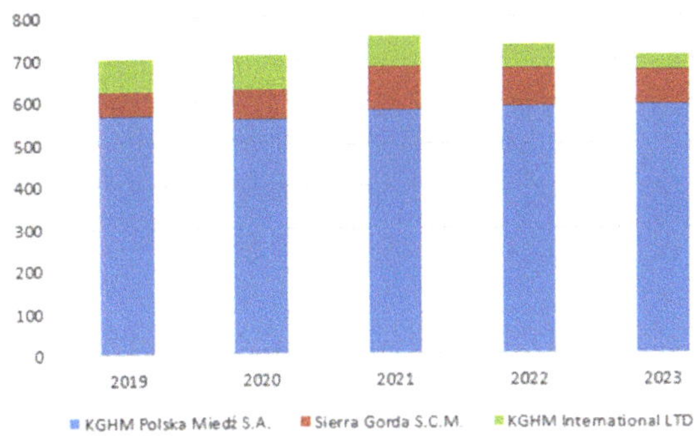

Figure 12. Copper production by KGHM Polska Miedź S.A. in 2019–2023 [thousand tons].

The volume of domestic production, amounting to 592,000 tons in 2023 (Figure 12), placed Poland in ninth position globally (Figure 13). At the time, global refined copper production was 27 million tons, and China alone produced as much as 45% (12 million t) of Cu. However, if domestic and foreign copper production were taken into account, it would rank seventh, just after the United States.

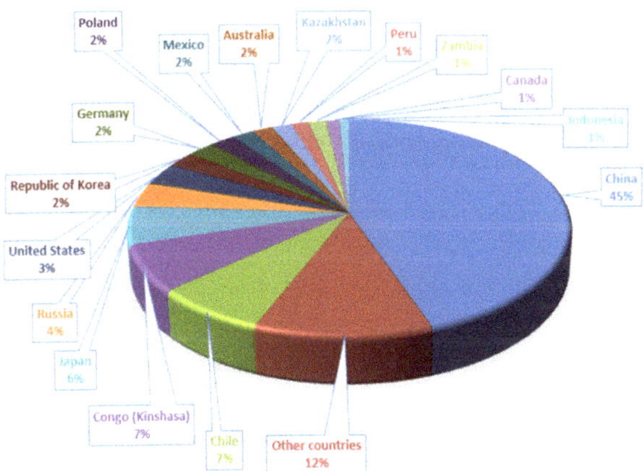

Figure 13. Copper refining.

The main product manufactured and sold by KGHM's Polish divisions is electrolytic copper (99.99% Cu), which is produced after the refining stage. At the processing stage, KGHM processes 49% of the electrolytic copper at HM Cedynia into other products desired in the market, i.e., copper wire rod, oxygen-free copper wire, low-alloy oxygen-free copper wire with silver, and copper pellets. Domestic concentrates are sold rather sporadically. It is rather the domestic subsidiaries that purchase concentrates to optimize capacity utilization at smelters. The copper mined in KGHM's foreign companies is mainly sold in the form of concentrates.

KGHM sells its copper products mainly on the European market, i.e., Poland (24%), Germany (21%), the Czech Republic (8%), Italy (7%), and Hungary (5%). About 10% of KGHM's copper also goes to China, where 60% of global copper was consumed in 2023 [37].

Figure 14 shows how capital expenditures in the KGHM Group evolved between 2022 and 2023. These expenses are divided into four categories, but the money for exploration and development is included in the development projects category. In each year, they accounted for more than 20% of expenditure. Compared to 2022, in 2023, spending on development projects increased by 5 pp. In 2022, it was about 662 million, and in 2023 it was almost 800 million, an increase of more than 20%. These figures indicate that the Polish KGHM Group capital expenditure does not lag behind the leading copper producers, but is among the highest.

The chart also shows the structure of development projects. KGHM spends most on the deposit access program (75%), and 19% is spent on the costs of exploration of the resource base.

Domestically, in order to secure future production, KGHM is working to explore copper ore deposits in the Radwanice and Retków areas. In 2023, the company also expanded the scope of operations within the Głogów Głęboki-Przemysłowy mining area.

KGHM International LTD is working on the Victoria Project in Ontario, Canada, where a new underground mine is planned. In 2002, mineral rights were acquired, exploration work began, and at the time of writing (2024), preparatory work is still underway. The lifetime of the mine is estimated at 16 years, with a projected annual copper production of 17,000 t Cu.

Regarding the Sierra Gorda development in 2023, 54% of expenses were related to the removal of overburden in order to make more portions of the deposit available for mining, and the remainder were related to the development and restoration of property, plant and equipment [38].

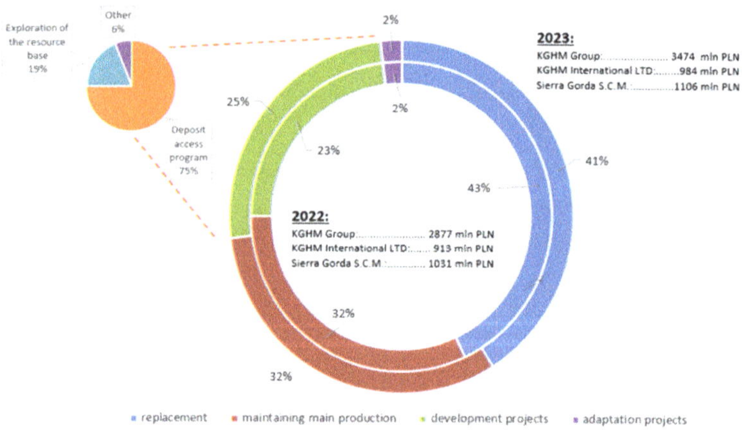

Figure 14. Group capital expenditure in 2022–2023.

In addition to the above new investments, KGHM International's assets include a copper-bearing deposit in British Columbia, Canada. An open pit mine is planned there and is known as the Ajax Project. Based on exploratory work, it was estimated that annual production could be 53,000 t Cu, and the mine itself could operate for 19 years. Unfortunately, as of now (2024), the project has not received an EA Certificate, so the Canadian government has not issued an operating permit.

Nevertheless, the scale of demand for copper remains enormous. The growth in mining production capacity is estimated at an average of 4.9% annually [38], which still may not be sufficient to meet the increasing demand. The ETC estimates that 25–30 new mines should be developed by 2050. However, it should be noted that the average time to complete such projects is approximately 17 years, considering the phases of exploration, obtaining licenses, and constructing mining facilities [40].

The KGHM Group has included two recycling-related goals in its strategy for 2030 titled "Strategy of the KGHM Group to 2030 with an outlook to 2040" [46]. In HM Legnica and HM Głogów, in 2022, copper scrap accounted for 22% of electrolytic copper production and was 131 thousand t Cu. In 2023 there was an increase in the share of scrap in electrolytic copper production to 25% (i.e., 142.2 thousand t Cu). KGHM is clearly showing progress in terms of copper recycling. In order to meet its targets, KGHM is stepping up activities related to the purchase of copper-bearing scrap from the so-called urban mining.

The company is also working to increase the efficiency of its copper recycling process. A high-grade copper scrap melting unit has been constructed in HM Legnica's process line. Ultimately, it will be supplemented with a scrap trading base, where feedstock will be prepared for smelting at HM Legnica and HM Głogów. In 2023, 83 thousand t of scrap metal was processed at HM Legnica (including 26 thousand t in the melting and refining furnace), which was thus recycled and would not contribute to landfill waste. Among the scrap metal recycled in 2023 at HM Legnica's facilities, 1,876 t comprised returns from other KGHM divisions [38].

In addition, through landfilling, KGHM disposes of flotation tailings, which account for 94% of the mined ore, or 28 million t per year [37]. Since 1974, flotation tailings have been deposited at the Żelazny Most Mining Waste Neutralization Facility (2100 hectares). At the beginning of KGHM's operation, the process of extracting copper from the ore was less efficient, and over time, the technology was improved, but this does not change the fact that considerable amounts of valuable elements, including copper, ended up in the Żelazny Most landfill. These elements still end up there, although their waste content has already been significantly reduced. However, KGHM is developing a plan to "explore and implement solutions to utilize landfills as secondary deposits" [38].

5. Conclusions

The energy transition is a major challenge for the entire world. One of the biggest threats is the supply of critical raw materials, without which the planned transformation of the world's shift to renewables and greening will not take place. It is estimated that demand for critical raw materials will be much higher than the level that is currently mined or recycled.

One of the most desirable raw materials is copper, the value chain of which is closely analysed in terms of the risks that are evident from the production of intermediates at each stage of value addition. Likewise, the margins generated at the respective stages of the production of copper intermediates and financial products are also analysed.

The value chain of copper produced for the energy transition and beyond is unstable. There is a high level of concentration in the scope of mining, with 70% mined by five countries: Chile, Peru, Congo, Zambia, and the USA. There is a similar concentration in refining, where China alone refines 45% of copper, which constitutes a significant threat. This high level of concentration poses a risk to supply stability, where all it takes is one political conflict or bad weather conditions for a significant supply stream to be curtailed and thus copper prices to skyrocket.

Future demand for copper exceeds the actual current supply, and recycling levels are insufficient. Europe is heavily dependent on the "big players" in all of this, and it takes at least 17 years to launch new copper mines.

According to studies, the number of newly discovered deposits declined between 2004 and 2023, despite increasing exploration expenditures. All this makes it possible to conclude that the salvation for Europe is the Polish company KGHM Polska Miedź S.A., which has the resources and ambitions stay among the top copper producers in the world.

This is where the main investment streams from the EU should go. Therefore, the authors of this study focused on presenting the case of KGHM Polska Miedź S.A. in the context of global supply chains, thereby emphasizing the importance of this company in the EU's energy transition.

On the one hand, there are scientific units where research work is being carried out on basic principles to reduce planned copper consumption. Work is underway to reduce the use of copper in renewable energy installations and EVs, which could cause the cumulative demand to fall by 30%. On the other hand, there are the mining and smelting plants, including KGHM's plants, which are keeping up with global trends, where investments are being made and work is being conducted to identify deposits, increase mining efficiency and increase the efficiency of the refining process.

At the same time, it is planned to raise the level of recovery of recycled copper (urban mining) from 60% to at least 90%, but a significant increase in recycling will not be possible until after 2030, when the copper involved in the transformation process will begin to return to circulation. It is estimated that more than 7 MT of copper can be obtained by 2050, which should meet more than 40% of the total demand for the energy transition. This means that appropriate regulations and economic incentives must be developed to make this assumption a reality. KGHM is prepared to accept copper from urban mining, while working to increase its capacity in this area.

Author Contributions: Conceptualization, A.K., M.P.-S., S.L.-S. and A.W.-S.; methodology, M.P.-S. and S.L.-S.; resources, A.K., M.P.-S., S.L.-S. and A.W.-S.; data curation, A.K., M.P.-S., S.L.-S. and A.W.-S.; writing—original draft preparation, A.K., M.P.-S., S.L.-S. and A.W.-S.; writing—review and editing, A.K.; visualization, A.W.-S.; All authors have read and agreed to the published version of the manuscript.

Funding: The authors wish to thank the Faculty of Civil Engineering and Resource Management at the AGH University of Krakow for the financial support for this research (no. 16.16.100.215).

Data Availability Statement: The original contributions presented in this study are included in the above article; further inquiries can be directed to the corresponding author.

Conflicts of Interest: The authors declare no conflicts of interest.

References

1. European Commission. Communication from the Commission to the European Parliament, the European Council, the Council, the European Economic and Social Committee and the Committee of the Regions. In *The European Green Deal, COM/2019/640 Final*; European Commission: Brussels, Belgium, 2019. Available online: https://eur-lex.europa.eu/legal-content/EN/TXT/HTML/?uri=CELEX:52019DC0640 (accessed on 16 July 2024).
2. Xu, Y.; Liu, A.; Li, Z.; Li, J.; Xiong, J.; Fan, P. Review of Green Supply-Chain Management Diffusion in the Context of Energy Transformation. *Energies* **2023**, *16*, 686. [CrossRef]
3. Genc, T.S.; Kosempel, S. Energy Transition and the Economy: A Review Article. *Energies* **2023**, *16*, 2965. [CrossRef]
4. IEA. *World Energy Outlook 2023*; Organisation for Economic Co-Operation and Development (OECD): Paris, France, 2023. [CrossRef]
5. Rietveld, E.; Bastein, T.; van Leeuwen, T.; Wieclawska, S.; Bonenkamp, N.; Peck, D.; Klebba, M.; Le Mouel, M.; Poitiers, N. *Strengthening the Security of Supply of Products Containing Critical Raw Materials for the Green Transition and Decarbonisation*; European Parliament: Strasbourg, France, 2022. Available online: http://www.airaassociazione.it/wp-content/uploads/2023/01/Strengthening-the-security-of-supply-of-products-containing-CRMs-for-the-green-transition-and-decarbonisation.pdf (accessed on 11 July 2024).
6. Hund, K.; la Porta, D.; Fabregas, T.P.; Laing, T.; Drexhage, J. *Minerals for Climate Action: The Mineral Intensity of the Clean Energy Transition*; International Bank for Reconstruction and Development, The World Bank: Washington, DC, USA, 2020. Available online: https://pubdocs.worldbank.org/en/961711588875536384/Minerals-for-Climate-Action-The-Mineral-Intensity-of-the-Clean-Energy-Transition.pdf (accessed on 11 July 2024).

7. Bobba, S.; Carrara, S.; Huisman, J.; Mathieux, F.; Pavel, C. Critical Raw Materials for Strategic; Technologies and Sectors in the EU—A Foresight Study. 2020. Available online: https://rmis.jrc.ec.europa.eu/uploads/CRMs_for_Strategic_Technologies_and_Sectors_in_the_EU_2020.pdf (accessed on 11 July 2024).
8. Zhang, Z.; Chen, S.; Zhang, H.; Yao, C.; Lou, H.; Chen, M.; Xia, B.; Sun, Y.; Lang, X.; Cai, K. Enhanced oxygen reduction kinetics of SrCoO3-δ by Ta/Cu or Nb/Cu co-doping as high-performance cathodes for SOFC. *Ceram. Int.* **2023**, *49*, 39003–39012. [CrossRef]
9. Quan, B.; Wang, Y.; Chen, Y.; Lu, X.; Teng, S.; Zhou, X.; Qiao, M.; Lai, H.; Huang, X. Manipulation of nano-metals to implement rational conduction tailoring for high-efficiency microwave absorption. *Carbon* **2023**, *210*, 118045. [CrossRef]
10. Das, S.; Ali, A.; Jana, R.N. Insight into the dynamics of magneto-casson hybrid nanoliquid caused by a plate rotation. *World J. Eng.* **2021**, *18*, 66–84. [CrossRef]
11. Wu, Y.; Lv, J.; Xie, F.; An, R.; Zhang, J.; Huang, H.; Shen, Z.; Jiang, L.; Xu, M.; Yao, Q.; et al. Single and double transition metal atoms doped graphdiyne for highly efficient electrocatalytic reduction of nitric oxide to ammonia. *J. Colloid Interface Sci.* **2024**, *656*, 155–167. [CrossRef]
12. Toyooka, G.; Tanaka, T.; Kitayama, K.; Kobayashi, N.; Watanabe, T.; Fujita, K.-I. Hydrogen production from cellulose catalyzed by an iridium complex in ionic liquid under mild conditions. *Catal. Sci. Technol.* **2021**, *11*, 2273–2279. [CrossRef]
13. Guo, F.; Qiao, Q.; Mao, S.; Bai, J.; Dong, K.; Shu, R.; Xu, L.; Wei, H.; Qian, L.; Wang, Y. A comprehensive study on the pyrolysis behavior of pine sawdust catalyzed by different metal ions under conventional and microwave heating conditions. *Energy* **2023**, *272*, 127115. [CrossRef]
14. Chen, H.; Hsu, H.-Y.; Hsieh, J.-E.Y.; Hung, H.-E. A differential privacy-preserving federated learning scheme with predictive maintenance of wind turbines based on deep learning for feature compression and anomaly detection with state assessment. *J. Mech. Sci. Technol.* **2024**, *38*, 3413–3429. [CrossRef]
15. Kushnir, V.; Koshkin, I.; Ibragimova, S. Electric Heating of Non-conductive Dispersed Raw Materials in Activated Carbon Production. In *Lecture Notes in Mechanical Engineering, Proceedings of the 6th International Conference on Industrial Engineering ICIE 2020, Sochi, Russia, 18–22 May 2020*; Radionov, A.A., Gasiyarov, V.R., Eds.; Springer: Berlin/Heidelberg, Germany, 2021. [CrossRef]
16. Yang, M.; Li, L.; Shi, J.; Xia, H.; Xu, J. Electrocatalytic hydrogenation of lignin-derived compounds in heteropolyacid fluidized system. *Biomass Convers. Biorefinery* **2024**, 1–11. [CrossRef]
17. Khrunina, N.P.; Cheban, A.; Yu, A. A theoretical approach to disintegration equipment design for mineral raw materials. *Eurasian Min.* **2023**, *40*, 66–69. [CrossRef]
18. Li, D.; Shen, L.; Zhong, S.; Elshkaki, A.; Li, X. Spatial and temporal evolution patterns of material, energy and carbon emission nexus for power generation infrastructure in China. *Resour. Conserv. Recycl.* **2023**, *190*, 106775. [CrossRef]
19. Li, H.; Yao, X.; Tachega, M.A.; Ahmed, D.; Ismaail, M.G.A. Technology selection for hydrogen production in China by integrating emergy into life cycle. *J. Clean. Prod.* **2021**, *294*, 126303. [CrossRef]
20. Eheliyagoda, D.; Wei, F.; Shan, G.; Albalghiti, E.; Zeng, X.; Li, J. Examining the temporal demand and sustainability of copper in China. *Environ. Sci. Technol.* **2019**, *53*, 13812–13821. [CrossRef]
21. Guo, J.; Zhu, K.; Cheng, Y. Deployment of clean energy technologies towards carbon neutrality under resource constraints. *Energy* **2024**, *295*, 131012. [CrossRef]
22. Wang, Y.; Guo, C.H.; Zhuang, S.R.; Chen, X.J.; Jia, L.Q.; Chen, Z.Y.; Xia, Z.L.; Wu, Z. Major contribution to carbon neutrality by China's geosciences and geological technologies. *China Geol.* **2021**, *4*, 329–352. [CrossRef]
23. Li, Y.; Pan, S.; Ning, S.; Shao, L.; Jing, Z.; Wang, Z. Coal measure metallogeny: Metallogenic system and implication for resource and environment. *Sci. China Earth Sci.* **2022**, *65*, 1211–1228. [CrossRef]
24. Li, L.; Tang, Q.-Y.; Zhang, Y. Comparative Study of Magmatic Sulfide Type Nickel-Cobalt Deposits in Cratons and Orogenic Belts—A Case Study of Jinchuan and Xiarihamu Deposits. In Proceedings of the 11th Academic Conference of Geology Resource Management and Sustainable Development, Beijing, China, 16–17 December 2023; Volume 1, pp. 426–432. [CrossRef]
25. Jia, J.; Xin, L.; Lu, C.; Wu, B.; Zhong, Y. China's CO_2 emissions: A systematical decomposition concurrently from multi-sectors and multi-stages since 1980 by an extended logarithmic mean divisia index. *Energy Strategy Rev.* **2023**, *49*, 101141. [CrossRef]
26. Domaracka, L.; Matuskova, S.; Tausova, M.; Senova, A.; Kowal, B. Efficient Use of Critical Raw Materials for Optimal Resource Management in EU Countries. *Sustainability* **2022**, *14*, 6554. [CrossRef]
27. Jarosiński, M.; Krzyœcin, J.; Ustrnul, Z.; Kochanek, K.; Janik, T. Geophysical research in solving civilization problems relevant to Poland. *Przegląd Górniczy* **2023**, *71*, 418–429.
28. Bhowmik, S.; Chaudhary, A.; Patra, M.; Patra, R. Oxygen reduction reaction by metallocorroles and metallophthalocyanines. In *Oxygen Reduction Reaction: Fundamentals, Materials, and Applications*; Elsevier: Amsterdam, The Netherlands, 2022. [CrossRef]

29. Khan, M.R.; Alqahtani, A.M.; Alhazmi, S.E.; Elkotb, M.A.; Sidi, M.O.; Alrihieli, H.F.; Tag-Eldin, E.; Yassen, M.F. Numerical Investigation of Darcy–Forchheimer Hybrid Nanofluid Flow with Energy Transfer over a Spinning Fluctuating Disk under the Influence of Chemical Reaction and Heat Source. *Micromachines* **2023**, *14*, 48. [CrossRef] [PubMed]
30. Memon, S.A.; Shaikh, H.I.; Raza, R.; Mughal, Z.U.N.; Memon, A.A.; Memon, S. Graphene incorporated mesoporous perovskite with excellent conductivity and catalytic activity for low temperature solid oxide fuel cells. *New J. Chem.* **2022**, *46*, 12530–12539. [CrossRef]
31. Kim, S.H.; Lee, D.H. Polyetherimide for Magnet Wire Applications. *SAE Int. J. Adv. Curr. Pr. Mobil.* **2022**, *5*, 517–520. [CrossRef]
32. ETC. Material and Resources Requirements for the Energy Transition. Available online: https://www.energy-transitions.org/publications/material-and-resource-energy-transition/ (accessed on 9 July 2024).
33. EY. Critical Raw Materials for the Energy Transition—How to Achieve the Targets. Available online: https://www.ey.com/en_bg/insights/energy-resources/critical-raw-materials-for-energy-transition (accessed on 4 July 2024).
34. U.S. Geological Survey 2024. Available online: https://pubs.usgs.gov/publication/mcs2024 (accessed on 6 July 2024).
35. U.S. Geological Survey 2013. Available online: https://pubs.usgs.gov/publication/mineral2013 (accessed on 6 July 2024).
36. U.S. Geological Survey 2023. Available online: https://www.usgs.gov/publications/mineral-commodity-summaries-2023 (accessed on 6 July 2024).
37. Raport Zintegrowany KGHM. Available online: https://raportcsr2022.kghm.com/ (accessed on 8 July 2024).
38. Sprawozdanie Zarządu KGHM. Available online: https://kghm.com/sites/default/files/2024-08/Sprawozdanie%20Zarz%C4%85du%20z%20dzia%C5%82alno%C5%9Bci%20Grupy%20Kapita%C5%82owej%20KGHM%20Polska%20Mied%C5%BA%20S.A.%20w%20I%20p%C3%B3%C5%82roczu%202024.pdf (accessed on 6 July 2024).
39. Rizos, V.; Bryhn, J.; Alessi, M.; Campmas, A.; Zarra, A. *Identifying the Impact of the Circular Economy on the Fast-Moving Consumer Goods Industry*; The European Economic and Social Committee: Bruxelles, Belgium, 2019.
40. IEA. The Role of Critical Minerals in Clean Energy Transitions. Available online: https://www.iea.org/reports/the-role-of-critical-minerals-in-clean-energy-transitions (accessed on 10 July 2024).
41. World Integrated Trade Solution. World Bank. Available online: https://wits.worldbank.org/ (accessed on 18 September 2024).
42. Fraunhofer Institute for Systems and Innovation Research ISI. Available online: https://www.isi.fraunhofer.de/en.html (accessed on 23 September 2024).
43. Directive (EU) 2023/2413 of the European Parliament and of the Council. Available online: https://eur-lex.europa.eu/legal-content/EN/TXT/HTML/?uri=OJ:L_202302413 (accessed on 8 July 2024).
44. Directive (EU) 2023/1791 of the European Parliament and of the Council. Available online: https://eur-lex.europa.eu/legal-content/EN/TXT/HTML/?uri=CELEX:32023L1791 (accessed on 8 July 2024).
45. Regulation (EU) 2024/1252 of the European Parliament and of the Council of 11 April 2024 Establishing a Framework for Ensuring a Secure and Sustainable Supply of Critical Raw Materials and Amending Regulations (EU) No 168/2013, (EU) 2018/858, (EU) 2018/1724 and (EU) 2019/1020 Text with EEA Relevance. Available online: https://eur-lex.europa.eu/legal-content/EN/TXT/HTML/?uri=OJ:L_202401252 (accessed on 16 July 2024).
46. Strategy of the KGHM Group to 2030 with an Outlook to 2040. Available online: https://kghm.com/en/strategy-kghm-group-2030-outlook-2040 (accessed on 10 July 2024).

energies

MDPI

Article

Modeling Energetic Effectiveness and Breakage Intensity in Mining and Processing Circuits on Limestone Crushing in HPGR

Agnieszka Saramak and Daniel Saramak *

Department of Environmental Engineering, Faculty of Civil Engineering and Raw Materials Economy, AGH University of Krakow, 30-059 Krakow, Poland; saramak@agh.edu.pl
* Correspondence: dsaramak@agh.edu.pl; Tel.: +48-12-617-4559

Abstract: This article refers to the aspects of energy consumption and comminution effectiveness in the mineral processing sector through the evaluation of limestone crushing in a high-pressure grinding roll. The investigative program included a series of crushing tests on limestone samples in a laboratory High Pressure Grinding Rolls (HPGR) press device. The tests were carried out in the scheme of factorial experiment with three levels of pressure (Fsp) and three levels of material moisture (M). The major finding was to determine energetic models referring to consumption of energy and reduction in Bond work index Wi, designed as a function of operational pressure in HPGR and material moisture. The other investigative results encompassed models on fineness effectiveness and throughput. The models appeared statistically significant and showed relationships both with pressure and moisture. The results of the investigations showed that Bond work index Wi decreases when the Fsp increases, but Wi increases as the moisture content decreases. The calculated models also showed an increase in unit energy consumption in the press together with increasing of Fsp and moisture. The models for throughput and finest particle content in HPGR product showed in turn that increasing of Fsp and M results in decreasing of the productivity.

Keywords: energy; HPGR; limestone; Bond work index; mineral processing; mathematical modeling

1. Introduction

The mining and metallurgical sectors play crucial roles in the technological development of the economy both on the domestic and world scale. However, technological circuits of raw material mining and processing consume a significant amount of energy, which may reach as much as 30% of total costs [1]. Processes of industrial comminution—a major part of raw material enrichment stage—are the most energy-consuming operations in the metal production value chain. An additional issue related to operations of crushing and grinding is their low breakage effectiveness. According to Legendre [2], only 10% of energy input is directly consumed on the feed material breakage, while the remaining portion is dissipated in the form of heat and noise [3]. In the industry of limestone powder production, the grinding operations are of key significance in entire circuits, and the energy consumption in operations of industrial comminution consume as high as 2% of world energy [1].

High-pressure grinding is technology with proved energy benefits both in ore processing circuits and in limestone grinding. The promising results concerned mostly the achieved energy savings and good material breakage results were demonstrated in numerous investigations [4–7]. However, it appears that a proper selection of operational parameters of the press device, determination the feed material characteristics, and adjustment of process course can also be of a key-significance in improving an effectiveness the HPGR operation. The press parameters and a process route should be verified towards the proper characterization of feed material, which mostly relates to a particle size, shape, moisture, and mineralization in case of ores [8–10].

As mentioned the above, various studies [8–14] have been carried out to evaluate the impact of the operational pressing force on comminution results in HPGR, regardless of the type of mining industry. They mostly referred to determining energy consumption and crushing ratio [11–14]. Investigations on the moisture content are less common; however, their results suggest that, in general, the HPGR feed with higher moisture shows a tendency to higher comminution intensity, while energy utilization rises as the moisture content of the feed increases. Investigations carried out for olivine sand [10] revealed that crushing ratio decreases for higher moisture, especially when the material is crushed at greater pressing force. Nevertheless, HPGR machines should not be considered a universal or flawless solutions for comminution of all types of materials. However, they might lead to achievement of technological and economic profits in comparison to conventional crushers, through achieving higher breakage intensity of material and lower unit energy consumption.

Results of the experiments need to be developed to formulate more general laws or models. Mathematical and statistical tools and methods can be helpful in the processing of these raw results. The methodology used in the paper includes regressive models based on the theory of multiple regression (linear and non-linear), correlation, and assumptions of mathematical modeling [15,16].

2. Significance and Scope

A significant number of investigations on HPGR were carried out in the hard rock industry, including iron and non-metallic ores. Considerable energy savings have been proven, which may reach as much as 30% or even higher, depending on the type of material and its comminution resistance [17,18]. However, the first industrial applications of HPGR devices in technological circuits of raw materials processing concerned the cement industry and limestone production. Energy savings in these circuits were proven especially in the downstream grinding stage in ball mills [19,20]. In the work [8], the potential for the optimization of HPGR operation were shown, from the scope of control the comminution ratio of HPGR products. It appeared that together with increasing the breakage intensity in HPGR (S_{80}), the energy consumption of the entire circuit of crushing and grinding was reduced for even 50%. Results of such investigations are very important, especially for the reason that cement production circuits and limestone grinding operations are based on operation of intense grinding of material in various types of mills. These devices consume huge amounts of energy and even a minor improvement of their operation may result in considerable saving on energy costs. There are many operational parameters of the HPGR device that influence the technological, economic, and energetic effects achieved in the press. The main purpose of the paper was to investigate the impact of pressure and moisture and such parameters like the speed of rolls, gap, and particle size of the feed were not analyzed. However, this review works on HPGR operation in relation to this problem [21,22]

The problem presented in this paper is also consistent with the global approach to better utilizing energy and may be potentially helpful in strategizing transformations of the mining and mineral processing sectors in terms of energy supply [23]. Such actions involve effective management of energy, including stable supply and storage, as well as taking advantage of sources of renewable energy [24,25]. It should also be noted that the approach presented in the paper can be found in the literature but applied to ores [26], while for limestone, such models were not used.

3. Materials and Methods

The purpose of the paper was to carry out a series of crushing tests on limestone in the laboratory-scale HPGR device (Figure 1) which was equipped with two rolls with plain surface of a diameter 0.3 m and roll width 0.1 m.

Figure 1. Laboratory HPGR device: (**a**) motor, (**b**) floating roll, (**c**) fixed roll, (**d**) feed bin, (**e**) bin opening chute, (**f**) product collecting bin.

Limestone rocks were used in investigations. The material was crushed in laboratory jaw crusher prior HPGR experiments to the top size of 20 mm. About 100 kg of material was used in the investigations, and after jaw crusher treatment, the entire feed was homogenized and divided into representative samples in a riffle splitter. The particle size distribution of feed is presented in Figure 2. The characteristic of the feed was determined with the use of the laboratory sieve shaker (dry analysis).

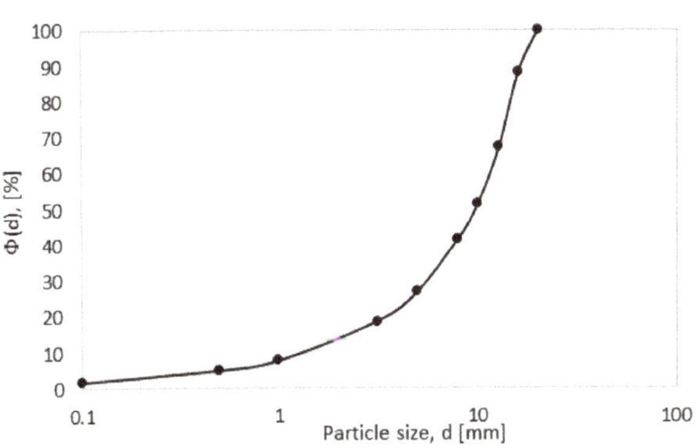

Figure 2. Particle size distribution of the feed material for HPGR tests.

The procedure of the investigative program was presented in Figure 3a (crushing test in HPGR machine) and Figure 3b (scheme for determining the Bond index Wi).

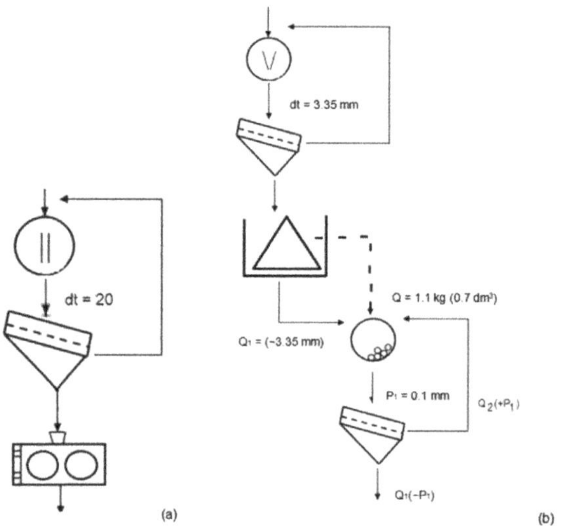

Figure 3. Research design: (**a**) crushing tests in HPGR, (**b**) scheme determination of the Bond work index Wi.

Research tests in HPGR testing consisted of nine individual tests at three different levels of the pressing force (F) and three values of material moisture (M). The original force F was recalculated on the pressure (Fsp), with values equaled to 3.3 N/mm^2, 4.3 N/mm^2, and 5.3 N/mm^2. The moisture values analyzed are as follows: 0, 2, and 4%. The gap-zero in HPGR was 4 mm, and the roll speed equaled 25 rpm (0.4 m/s).

The crushing product from the device was analyzed for the determination of the PSD and the Bond work index, according to the procedure developed by Bond [27,28]. The Bond Crushing Law states that the work required to obtain particles of size d from a feed of a greater size D is directly proportional to the square root of the surface-to-volume ratio of the product:

$$E[\mathrm{kWh/Mg}] = 10{\cdot}\mathrm{Wi}{\cdot}\left(\frac{1}{\sqrt{d_x}} - \frac{1}{\sqrt{D_x}}\right) \tag{1}$$

where D_x—characteristic particle of the feed (i.e., 80%), d_x—is the characteristic particle of the product (i.e., 80%), and Wi—Bond is the grindability index [27,28].

The sample for Bond procedure was prepared in a laboratory crusher (Figure 3b). A maximum amount the crushing product's particles were passing through the closing sieve 3.35 mm. After crushing the entire material, the sample was homogenized, and approximately 0.7 dm^3 of the material, which corresponds to about 1.1 kg, was taken to the first cycle of grinding. In the figure, this is represented by dashed line. Specific grinding tests were carried out in the laboratory via Bond's ball mill with the same diameter and length of 305 mm, filled out with 20.1 kg of steel balls grinding media. The grinders had different diameters that varied from 15.2 to 38.1 mm. The material mass lower than 0.1 mm, Q1, was removed, after each grinding cycle, and the remaining material, Q2, containing particles greater than 0.1 mm, was refilled with a portion of material Q1, taken from the feed. In the figure, it was marked with a solid-lined arrow that ends in the ball mill and is described "Q1 = −3.35mm". The symbol ($-P_1$) denotes the portion of material finer than 0.1 mm that was removed after each milling cycle, while (+P_1) denotes the portion of material greater than 0.1 mm that is returned to the next cycle of grinding.

Raw results for individual crushing tests and registered productivity and energy indices are presented in Table 1.

Table 1. Results of experiments.

Pressure, Fsp [N/mm^2]	Moisture Content, M [%]	Throughput, Q [Mg/h]	Specific Energy, Esp [kWh/Mg]
3.3	0	3.47	2.24
3.3	2	3.40	2.33
3.3	4	3.27	2.46
4.3	0	3.33	2.82
4.3	2	3.28	2.93
4.3	4	3.16	3.22
5.3	0	3.19	3.30
5.3	2	3.14	3.41
5.3	4	3.08	3.82

Throughput of the press, Q, and the specific energy consumption, E_{sp}, for individual experiments were calculated from Equations (2) and (3).

$$Q[\mathrm{Mg/h}] = \frac{m[\mathrm{kg}] \cdot 1000}{t[\mathrm{s}] \cdot 3600} \tag{2}$$

and

$$E_{sp}[\mathrm{kWh/Mg}] = \frac{P[\mathrm{kW}]}{Q[\mathrm{Mg/h}]} \tag{3}$$

where m—mass of a sample, t—crushing test duration, Q—capacity, and P—power of a motor.

The obtained crushing data were used in the determination of a series of mathematical models based on the specific pressing force and the feed moisture as independent variables. The values (dependent variables) that were the subject of the modeling are as follows:

- Bond work index Wi,
- Specific energy consumption Esp,
- Throughput Q,
- Crushing intensity expressed through crushing ratios,
- Generation of fines below 0.1 mm in HPGR product.

The statistical significance and potential autocorrelation of all coefficients and variables in the calculated models were checked on the probability level $(1 - \alpha) = 95\%$. The r-Pearson's linear correlation coefficient r [16] was used in the determination of the relationships between pairs of specific variables, and the non-parametric statistical test, based on the t-Student theoretical distribution, was used for verification of statistical significance for all r coefficients. When a given coefficient is significant, it means that there exists a correlation between the specific parameters.

4. Results of Investigations

4.1. HPGR Products Evaluation

In the first stage of analysis, a general fineness of HPGR products in terms of achieved crushing ratios was analyzed. Specific calculations were carried out using Equation (4) [29].

$$S_x = \frac{D_x}{d_x} \tag{4}$$

where x—the characteristic particle, D_x—particles in the feed: average (D_{50}) or eighty percentage (D_{80}), and d_x—particles in the product (d_{50} and d_{80}). The crushing ratio is a relationship of the characteristic particle of the feed to the characteristic particle of the crushing product, which presents how many times the size of a crushed product was reduced compared to the size of feed material. The specific outcomes are shown in Table 2. The results indicate that the pressure in HPGR devices generally had more of a significant impact on the breakage intensity of HPGR products than the moisture content. Additionally, alongside the increasing of Fsp, the breakage intensity increases as well, and a higher moisture content results in less significant decrease in values of crushing ratio. These relationships are true for all analyzed ratios. It may indicate that HPGR devices operating at higher pressure values produce fines more effectively than coarser particles.

Table 2. Results of experiments—achieved crushing ratios.

Pressure Fsp, [N/mm^2]	Moisture Content M, [%]	S_{20}	S_{50}	S_{80}
3.3	0	8.86	5.12	3.2
3.3	2	7.66	5.1	3.18
3.3	4	6.99	4.88	3.1
4.3	0	14.19	7.49	4.32
4.3	2	10.98	6.57	3.62
4.3	4	10.65	5.63	3.34
5.3	0	15.72	8.31	4.64
5.3	2	15.36	8.22	4.66
5.3	4	13.93	8.11	4.68

The above relationship was also confirmed through analysis of yield of finest particles (below 100 μm). The moisture content in the feed has an inverse impact on the yield of fines, because lower contents of finest particles can be observed in HPGR product with higher moisture content (Figure 4).

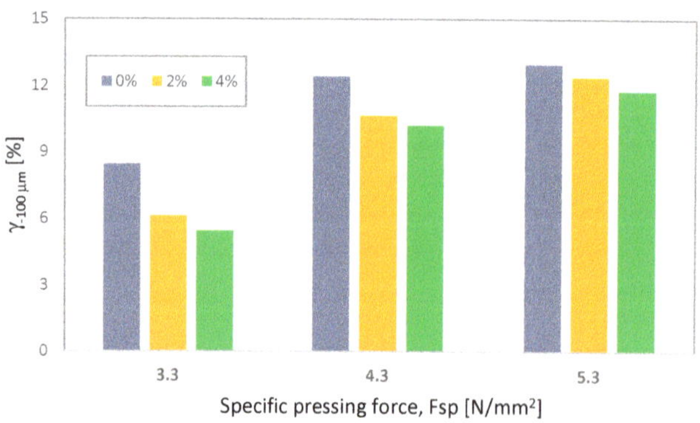

Figure 4. Yields of particles below 100 microns in HPGR products crushed under various pressure and with various moisture contents.

Table 1 shows throughput values achieved in individual experiments. These results indicate an inverse relationship between the press productivity and pressure because of the higher values of the pressing force throughput drops. A similar relationship can be observed between pressure and moisture: for higher moisture content throughput decreases, too. This can be caused through a greater slip of the material on the plain surface of both rolls. The lowest decrease in the Q value was observed with the highest moisture content.

4.2. Evaluation of Energetic Aspects

The primary purpose of investigations was to assess possible energy benefits obtained for variable pressure in HPGR and the material moisture, and for this purpose the Bond work indices Wi were calculated (Table 3).

$$W_i = \frac{16}{G^{0.82}} \cdot 1.1 \left[\frac{\text{kWh}}{\text{Mg}}\right] \quad (5)$$

Table 3. Bond work indices Wi obtained for all HPGR crushing products.

Pressure, Fsp [N/mm^2]	Moisture Content, M [%]		
	0%	2%	4%
3.3	10.92	10.44	10.18
4.3	10.53	10.22	10.01
5.3	10.44	10.18	9.97

Equation (5) is true for the control sieve 0.1 mm and characteristic particles expressed in millimeters [27,28].

The results obtained indicate that both the pressure, Fsp, and the material moisture, M, influence the Bond values Wi. As the Fsp increases, Bond indices drop less than proportionally, which may indicate that the performance of the HPGR device with too high of an operational pressing force may not necessarily be effective from energetic and economic scopes. However, it should be underlined that this is a rather complex issue because other parameters also have an impact on the obtained effects (i.e., comminution resistance of material, particle size composition, operational regime of the press like the rotational speed of rolls, gap width, stream of the recycled material, among others). In this case, it is evident that the moisture content positively impacts the reduction in the Bond Wi value, as an increase in M leads to the decrease in Wi, which could potentially result in energy savings. Alternatively, too much moisture in the material may cause a slippage of the particles on the surface of the rolls which leads to reduction in the HPGR throughput index. This phenomenon is very individual and depends on the material type, but for the tested material, it can be noticed that moisture 4% yet reduces Wi indices. The value of Bond work index for the feed material was 11.21 kWh/Mg.

Both Fsp and M have a proportional impact on increasing energy consumption. Figure 5 indicates that we can talk about a proportional or even linear relationship between the specific energy consumption and operational pressure Fsp, as well as between unit energy consumption and moisture content.

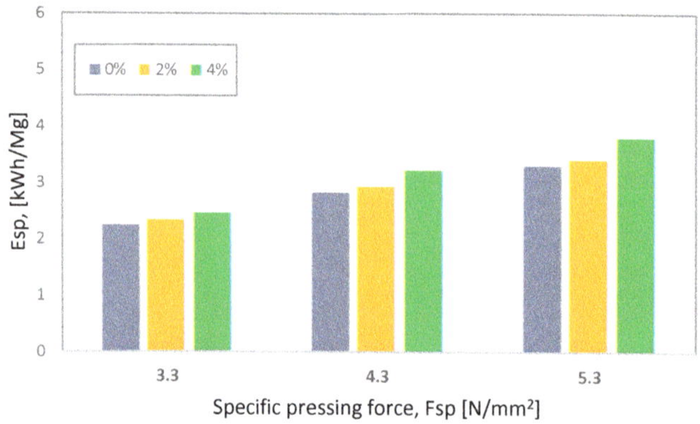

Figure 5. Specific energy consumption for individual HPGR products crushed under various pressure and with various moisture contents.

5. Mathematical Models of HPGR Operation

As was mentioned in the Section 3, the models of the HPGR press device operation that were worked out are as follows: a model for Bond work index, models for energy consumption and press throughput, technological models concerning breakage intensity and finest particles production. Two independent variables—pressure and feed moisture—were used as independent values in each model. It should be underlined that both these parameters have different units: Fsp is expressed in [N/mm^2] and moisture M in [%]. Therefore, all presented analyses apply to the following interpretation: a single unit of change for Fsp is 1 N/mm^2 and for M it is 1%. Prior to modeling, the correlation between specific variables were checked (Table 4). The results show that Fsp and M are not correlated.

Table 4. Calculated correlation matrix.

	Fsp	M	Wi	Esp	Q	γ_{-100}	S_{20}	S_{50}	S_{80}
Fsp	1.00	0.00	−0.46	0.94	−0.83	0.88	0.93	0.94	0.93
M	0.00	1.00	−0.84	0.31	−0.53	−0.33	−0.31	−0.23	−0.21

Values indicated in red (Table 4) are statistically significant at the accepted probability level 95%. The Fsp is significantly correlated with all the other variables. The above results are in line with individual models, i.e., the pressure has an impact to breakage intensity, production of finest size fractions, and savings of energy consumption in downstream grinding processes. The pressure is also correlated negatively with throughput, which is convergent with the general relationship observed in plant conditions. Value of the correlation coefficient between the pressing force and unit energy consumption is even higher (0.94) and this is also confirmed in investigations presented in reference [6]. The M parameter, in turn, is significantly correlated only with the Wi Bond index and productivity Q.

5.1. Models Connected with Energy Consumption

Equation (6) shows the relationship between Bond index, F_{sp} and M. For this type of material, both the pressure and the material moisture have a positive impact on potential energy savings in the HPGR press.

$$Wi = 11.3 - \underset{[0.04]}{0.16} \cdot F_{sp} - \underset{[0.02]}{0.14} \cdot M \tag{6}$$

where F_{sp}—pressure, [N/mm^2], *M*—material moisture, [%], and *Wi*—Bond index, [kWh/Mg]. The estimation errors of specific coefficients are presented in square brackets in a line right below the coefficient. These errors were listed in Equations (6)–(9). Bonds work indices had the lowest values for the highest pressure and moisture values. Models were determined by means of the multiple regression method and the Wi was defined as a function of the HPGR operational parameter and material properties. Initially, a multiple regression linear model with Fsp and M as independent variables, was built. Next, the M parameter in the second power (M^2) was tested, but it did not bring improvements in model quality. The determination coefficient R^2 equals 0.902. All the tested variables appeared statistically significant at the accepted level $(1 - \alpha) = 0.95$. The constant in the model has no practical interpretation.

The model indicates that increasing of Fsp by 1 N/mm^2 decreases Wi by 0.16 kWh/Mg, while increasing of M by 1% effectively decreases Wi by 0.14 kWh/Mg. It appears that for the tested material, the feed moisture is of potentially lower influence for decreasing the Bond work index, compared to Fsp. However, both tested parameters positively influence a decrease in values of Wi index and also possible saving of energy in downstream processes of crushing and grinding.

However, it is important to note that this model is only partially aligned with operational reality, specifically within a certain range of HPGR pressure and material moisture. As the pressing force in the device increases beyond this range, both the fineness and the reductions in Wi decrease progressively. In contrast, the moisture content behaves differently, as an excessive amount of water in the feed typically has a detrimental effect on the process. Too high moisture of the material negatively influences both the throughput, through the slip of the material on rolls, and insufficient breakage.

The energy model expressed by Equation (7), which is based on a specific unit energy consumption, also shows a significant relationship but generally with Fsp. The coefficient of model determination R^2 equals 0.970. The M parameter has some impact on energy consumption, but it is of much lower significance, compared to the pressure. The model better describes the real situation, compared to model (6), because only 3% of the variability is not explained by variables incorporated into the model function.

$$\begin{array}{c} Esp = 0.24 + \underset{[0.04]}{0.58} \cdot F_{sp} + \underset{[0.02]}{0.09} \cdot M \end{array} \tag{7}$$

5.2. *Technological Models of HPGR*

An impact of the selected parameters on the HPGR productivity and the yield of the finest size fractions in HPGR product, was also evaluated. Equation (8) shows the model for HPGR throughput.

$$Q = 3.87 - \underset{[0.01]}{0.13} \cdot F_{sp} - \underset{[0.005]}{0.04} \cdot M \tag{8}$$

The model demonstrates that both F_{sp} and material moisture negatively affect press throughput, with F_{sp} having a significantly greater impact than moisture content. The model's explanation of the phenomenon, as indicated by the R^2 value, is 96.5%.

The content of the finest particle size fractions is crucial in limestone flour production. If the HPGR product contains a significant yield of these fractions, they might be bypassed or separated from the material prior to downstream grinding, thus mills grind a lower amount of material and energy costs are lower, too. For this purpose, the yield of particle size fraction finer than 0.1 mm ($\gamma_{-0.1}$) was also modeled by Formula (9).

$$\gamma_{-100} = -30.85 - \underset{[0.38]}{1.64 \cdot F_{sp}^2} + \underset{[3.29]}{17.03 \cdot F_{sp}} - \underset{[0.11]}{0.54 \cdot M} \quad (9)$$

As the F_{sp} value increases, the yield of fines in the HPGR product also rises, with the relationship approached by a hyperbolic or parabolic shape, therefore the F_{sp} in the second power was included in the model. It has increased the model quality, because R^2 equals 0.963, while for the model without Fsp^2 equaled 0.855.

5.3. Models for S_x Crushing Ratios

A set of models describing comminution ratios Sx were determined. Finally, the models for S_{20}, S_{50}, and S_{80} comminution were calculated. Results were presented in Table 5.

The impact of Fsp on a value Sx in each model is positive, while M is negatively bound. The M value is insignificant in models for S_{50} and S_{80}, and Fsp is statistically significant in all models. The modeling results also indicate that pressure has the greatest effect on the comminution ratio in the S20 model, with its influence diminishing for successive characteristic particles (Sx). Overall, the impact of Fsp is consistently greater than the effect of M.

Table 5. Models for S_x comminution ratios as functions of F and M.

Characteristic Particle, x	Model	R^2
20	−2.73 + 3.58 Fsp − 0.6 M	0.953
50	0.10 + 1.59 Fsp − 0.19 M	0.928
80	0.78 + 0.75 Fsp − 0.09 M	0.871

The evaluation of model accuracy, represented by the R^2 index, reveals that all models explain the phenomenon with an accuracy ranging from 87% to 95%. The highest R^2 value was observed in the S20 models, while the lowest was found in the S80 models. Based on this, these models can be effectively used for accurate forecasting of comminution results, particularly for prediction of fines content in HPGR products.

6. Conclusions

As was presented in the previous chapters, the effects of HPGR are usually visible in downstream comminution operations, which mostly result in lowering of grinding energy consumption and increasing the yield of fines. These advantages, however, are a direct result of the breakage intensity in HPGR machine. Nevertheless, the HPGR pressure and moisture of the material influence the achieved technological and energetic results. For the material analyzed, the Bond work indices decreased as pressure and moisture increased, with this relationship being close to hyperbolic. Alternatively, for the yield of the finest particle size fractions, the pressure had a higher impact comparing to the moisture and this parameter was in a second power in the model. Models for S_x showed in turn that Fsp had a positive impact on breakage intensity, with the most favorable modeling results achieved for S_{20}.

Practical implementation of these models may be helpful in achieving better technological results, like the higher yield of the finest particles in crushing products and a more effective course of the grinding operation. This is related to the higher degree of breakage of the feed material. Presented models are helpful in more accurate prediction and control of particle size composition of the crushing products, but in the presented form they cannot be extrapolated beyond the range of plant operational conditions resulting from technological regimes. For example, too low of a pressing force in HPGR results in insufficient comminution, while too high of a moisture level affects the circuit's productivity and a breakage due to the material slip.

The main contribution of the paper is to show the potential impacts of the parameters investigated on selected technological and economic effects. The models are not unique for all types of analyzed material, but only values of specific coefficients should change for various tested materials. Thus, the functional form of the model should be the same, and only the coefficients need to be recalculated. It was confirmed in similar investigative programs, for example, for ores crushed with HPGR. On the other hand, more advanced programs may contribute to work out more general and more accurate models of HPGR operation.

Author Contributions: Conceptualization, A.S. and D.S.; methodology, A.S. and D.S.; software, A.S.; validation, A.S.; formal analysis, A.S. and D.S.; investigation, D.S.; data curation, A.S.; writing—original draft preparation, A.S. and D.S.; writing—review and editing, A.S.; visualization, D.S. All authors have read and agreed to the published version of the manuscript.

Funding: This research received no external funding.

Data Availability Statement: The original contributions presented in the study are included in the article, further inquiries can be directed to the corresponding author.

Acknowledgments: Article is an effect of realization of project 101091885 "A Holistic Digital Mine 4.0 Ecosystem", (acronym mine.io), financed through European Health and Digital Executive Agency within the frames of HORIZON Innovation Actions, action: HORIZON-CL4-2022-RESILIENCE-01-06.

Conflicts of Interest: The authors declare no conflicts of interest.

References

1. Tromans, D. Mineral comminution: Energy efficiency considerations. *Min. Eng.* **2008**, *21*, 613–620. [CrossRef]
2. Legendre, D.; Zevenhoven, R. Assessing the energy efficiency of a jaw crusher. *Energy* **2014**, *74*, 119–130. [CrossRef]
3. Morrell, S. Predicting the overall specific energy requirement of crushing, high pressure grinding roll and tumbling mill circuits. *Min. Eng.* **2009**, *22*, 544–549. [CrossRef]
4. Abouzeid, A.Z.; Fuerstenau, D.W. Grinding of mineral mixtures in high-pressure grinding rolls. *Int. J. Min. Proc.* **2009**, *93*, 59–65. [CrossRef]
5. Aminalroaya, A.; Pourghahramani, P. The Effect of Feed Characteristics on Particles Breakage and Weakening Behavior in High Pressure Grinding Rolls (HPGR). *Min. Proc.Extr. Met. Rev.* **2022**, *43*, 610–621. [CrossRef]
6. Atmaca, A.; Kanoglu, M. Reducing Energy consumption of a raw mill in cement industry. *Energy* **2012**, *42*, 261–269. [CrossRef]
7. Tavares, M. Particle weakening in high-pressure roll grinding. *Min. Eng.* **2005**, *18*, 651–657. [CrossRef]
8. Aydogan, N.M.; Ergun, L.; Benzer, H. High pressure grinding rolls (HPGR) applications in the cement industry. *Min. Eng.* **2006**, *19*, 130–139. [CrossRef]
9. Numbi, B.P.; Zhang, J.; Xia, X. Optimal energy management for a jaw crushing process in deep mines. *Energy* **2014**, *68*, 337–348. [CrossRef]
10. Saramak, D.; Kleiv, R.A. The effect of feed moisture on the comminution efficiency of HPGR circuits. *Min. Eng.* **2013**, *43–44*, 105–111. [CrossRef]
11. Fuerstenau, D.W.; Abouzeid, A.-Z.M. Role of feed moisture in high-pressure roll mill comminution. *Int. J. Min. Proc.* **2007**, *82*, 203–210. [CrossRef]
12. Torres, M.; Casali, M. A novel approach for the modelling of high-pressure grinding rolls. *Min. Eng.* **2009**, *22*, 1137–1146. [CrossRef]
13. Pamparana, G.; Klein, B. A methodology to predict the HPGR operational gap by using piston press tests. *Min. Eng.* **2021**, *166*, 106875. [CrossRef]
14. Benzer, H.; Aydogan, N.A.; Dündar, H. Investigation of the breakage of hard and soft components under high compression: HPGR application. *Min. Eng.* **2011**, *24*, 303–307. [CrossRef]

15. Eck, C.; Garcke, H.; Knabner, P. *Mathematical Modeling*; Springer: Cham, Switzerland, 2017; pp. 1–509.
16. Navidi, W.C. *Statistics for Engineers and Scientists*, 4th ed.; McGraw-Hill Education: New York, NY, USA, 2014.
17. Patzelt, N. High pressure grinding rolls, a survey of experience. In Proceedings of the IEEE Cement Industry Technical Conference, Dallas, TX, USA, 10–14 May 1992; Volume 10, p. 180.
18. Rule, C.M.; Minnaar, D.M.; Sauremann, G.M. HPGR—revolution in Platinum? *J. S. Afr. Inst. Min. Met.* **2009**, *108*, 23–30.
19. Kurdowski, W. State-of-the art in the domain of comminution. *Cem. Wapno Beton* **1998**, *3*, 87–92.
20. Kalinowski, W. Modernization of the grinding set in cement applications in the light of the results of using techniques. *Surowce Masz. Bud.* **2006**, *1*, 12–15. (In Polish)
21. Rodriguez, V.A.; Campos, T.M.; Barrios, G.K.P.; Tavares, L.M. Review of recent advances in HPGR design. *Min. Proc. Extr. Met. Rev.* **2024**, 1–18. [CrossRef]
22. Saramak, D.; Saramak, A. Review of high-pressure technology in terms of technological and economic benefits achieved in raw materials processing. *Min. Proc. Extr. Met. Rev.* **2024**, *45*, 991–1004. [CrossRef]
23. Carvalho, M.; Millar, D.L. Concept Development of Optimal Mine Site Energy Supply. *Energies* **2012**, *5*, 4726–4745. [CrossRef]
24. Pouresmaieli, M.; Ataei, M.; Qarahasanlou, A.N.; Ghiyasi, Z. The Strategy of using Renewable Energies in the Mining Sector. *Iran. J. Min. Eng.* **2024**, *19*, 39–61.
25. Zhironkin, S.; Dotsenko, E. Review of Transition from Mining 4.0 to 5.0 in Fossil Energy Sources Production. *Energies* **2023**, *16*, 5794. [CrossRef]
26. Saramak, D.; Leśniak, K. Impact of HPGR operational pressing force and material moisture on energy consumption and crushing product fineness in high-pressure grinding processes. *Energy* **2024**, *302*, 131908. [CrossRef]
27. Bond, F.C. Crushing and grinding calculations. Part I. *Brit. Chem. Eng.* **1961**, *6*, 378–385.
28. Bond, F.C. Crushing and grinding calculations. Part II. *Brit. Chem. Eng.* **1961**, *6*, 543–548.
29. Lowrison, G.C. *Crushing and Grinding: The Size Reduction of Solid Materials*; Butterworths: London, UK, 1974; pp. 1–286. ISBN 978-0408705745.

energies

MDPI

Article

Impact of Power Quality on the Efficiency of the Mining Process

Tomasz Siostrzonek [1,*], Jakub Wójcik [1], Mateusz Dutka [1] and Wojciech Siostrzonek [2]

1 Department of Power Electronics and Automation of Energy Transformation Systems, AGH University of Krakow, 30-059 Krakow, Poland; jakubwojcik@agh.edu.pl (J.W.); mdutka@agh.edu.pl (M.D.)
2 Bachelor Computer Science and Engineering, Faculty of Electrical Engineering, Mathematics & Computer Science, Delft University of Technology, 2628 CD Delft, The Netherlands; wojciechsiostrzonek@gmail.com
* Correspondence: tsios@agh.edu.pl

Abstract: There are currently more than 30 underground mines operating in Poland. These are mines extracting hard coal, salt, and metal ores. Each of these plants has its own specifics for operation, but all operate under the same regulations. The basic principle is to ensure the safety of the crew and equipment. The progressive mechanization and automation of the mining process results in the installation of power electronic converters in the networks of mining plants, which significantly deteriorate the quality of the power in the plant supply networks. In addition, the constant reconfiguration of these networks related to the progress of the work can affect the safety conditions of the plant. This article describes problems occurring at one underground mining plant that are related to the structure of the power grid. Failures and interruptions in the production process were the result of poor power quality. They directly translated into increased production costs and significantly affected the safety level of the workforce, which could result in further consequences, not only in the financial sphere. The article also addresses the issue of existing legal regulations, the provisions of which may be insufficient in assessing the current state of power quality in mining plants.

Keywords: power quality; energy consumption in mine; power quality measurements

1. Introduction

The problem of power quality occurs at any location where electricity flows, i.e., from generation through transmission and distribution to the end user [1–6]. Power quality began to be discussed a dozen years ago. However, until now, the topics related to it are not treated on a par with other operational problems in industrial plants. The significance of the problem is understood in light of the relatively new field, where connections between the failures that occur and the problems caused by poor quality of the power supply are scarcely noticed. A special type of industrial plant, in this regard, is an underground mine. Therefore, it is worthwhile to clarify what the quality of the electricity supply entails, what it is, and what the consequences of negligence in this area means for the proper operation of the mining plant.

Sources of higher harmonics in the power grids of mines are primarily power electronic converters supplying the drive systems of process equipment. Such equipment includes, among others:

1. hoisting machines,
2. fans, and
3. conveyor and pump drives.

Different types of converters belonging to one of the groups are used in these devices:

- AC/DC converters,
- compound converters—multi-pulse AC/DC systems,
- AC/DC/AC intermediate frequency converters, and

- soft starters.

The largest electricity loads in an underground mine are generated by the hoisting machine, and it is also the source of a great many disturbances. The problem is not just the unit power of such a machine, but its operating mode, characterized by numerous starts and stops. The power of the hoisting machines can reach values of several megawatts. The drives of main ventilation fans require much lower power, ranging in power from a few hundred kilowatts to 1.5 MW. A much greater risk is posed by equipment with smaller capacities that are switched on in a completely uncoordinated manner. In these cases, it is not possible to take preventive measures.

In addition, the situation is exacerbated by the permanent reconfiguration of the electricity network due to the progress of the work, which necessitates changes in cable lengths and the relocation of equipment.

2. The Definition of Power Quality

Throughout the structure in which the flow of energy occurs, different understandings of the characteristics of its quality emerge, which provokes conflicts between suppliers and consumers of energy. If energy quality is mentioned, it can have many meanings depending on who defines it. Figure 1 illustrates the different perceptions of the negative effects of a certain event.

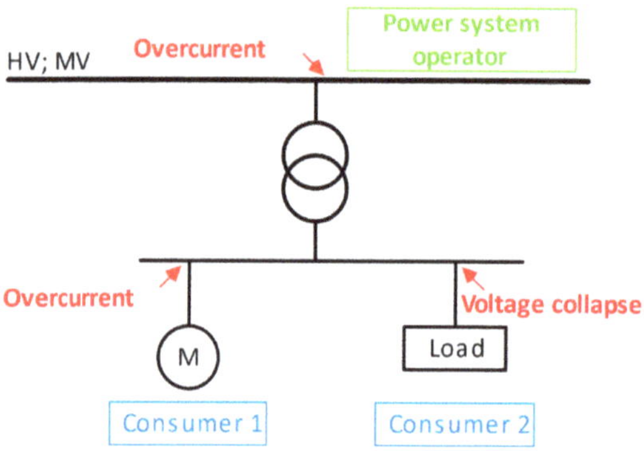

Figure 1. A perspective on power quality problems.

The first recipient's activation of one large drive results in overcurrent from his point of view. The electricity supplier will feel the same way. However, recipient 2 will have a completely different perception in this regard. For him, this procedure will be seen as a voltage collapse, a phenomenon unfavorable to him. Because the end customer experiences poor power quality and is affected by these phenomena, his perception of quality problems is of primary importance. Various definitions are cited in the literature, and this problem is still being discussed [7,8]. In [7], the definition that most captures the meaning of these discussions is provided below.

The quality of the power supply is a set of parameters describing the characteristics of the process of supplying power to the user under normal operating conditions, determining the continuity of the power supply (long and short interruptions) and characterizing the supply voltage (value, asymmetry, frequency, and time waveform).

Very often in mining plants, issues related to the quality of the power supply are incorrectly understood. It is wrongly believed that exceeding permissible levels of reactive power, and the associated penalties, is a problem related to the quality of the electricity supply. However, this is not true, as is clear from the cited definition.

3. The Assessment of the Power Quality

When determining the impact of equipment on the quality of the electricity supply, measurements of the relevant quantities are made. For this, class A measuring instruments are used in accordance with the IEC 61000-4-30 standard [9]. To assess the quality of the supply voltage, the following indicators are adopted:

RMS

Root mean square voltage is the equivalent direct current (DC) voltage for an alternating current (AC) source. The RMS voltage of a sinusoidal source of electromotive force is used to characterize the source. It is the square root of the time average of the voltage squared. The sinusoidal signal is defined as $u(t) = U_m \sin(\omega t)$, $U_{RMS} = U_m/\sqrt{2}$.

Frequency

This is the parameter describing the number of voltage waveform periods per second. According to the standard PN-IEC 60038, 50 Hz is the value considered correct. The acceptable frequency deviation is $\pm 10\%$

Total harmonic distortion (THD)

THD, which is defined as the percentage ratio of the sum of the voltage or current of all harmonic components to the voltage or current of the fundamental frequency, is a measurement of the harmonic distortion present in a signal.

Asymmetry

Supply voltage asymmetry refers to unevenness between the effective value of phase voltages or the shift angles between subsequent phases. In the case of three-phase systems, voltage asymmetry occurs when the effective values of three-phase voltages are not the same and/or the displacement angles between them are different than 120 degrees.

Fluctuation

The voltage fluctuations are defined as a series of changes in the RMS value or envelope of the voltage time course, the magnitude of which should not exceed $\pm 10\%$ of the rated network voltage. The main source of voltage fluctuations are devices with time-varying reactive power.

Distortion

A change in the supply voltage value is defined as an increase or decrease in the voltage value, usually caused by a change in the load on the power grid. Thera are two types of voltage changes: slow changes called voltage deviations and fast changes called voltage fluctuations. Two parameters are used, including short-term flicker indicator P_{st} and long-term flicker indicator P_{lt}.

Collapses and rises

Voltage collapse (sag) is a sudden reduction of the supply voltage below a specified level, typically -10% or sometimes -15%, and a subsequent return to the initial value or a value close to it. The duration of the dip is generally defined as a range from 10 ms to 1 s. The depth of the voltage dip is defined as the difference between the minimum effective value during the dip and the rated voltage.

Voltage surges are defined as sudden and short-term increases in voltage above 110% of the rated voltage.

Power outages

Power outages are situations in which the voltage at a power grid connection drops below 1% of the rated voltage. This parameter determines the reduction of the power supply voltage value below the level specified by the standard for a given recipient. A power outage can also be defined as a voltage drop to a value close to zero at the point of electricity supply. Power outages are classified according to length:

- Micro-breaks (transient)—last no longer then 1 s;
- Short—last longer than 1 s, but no longer than 3 min;
- Long—last over 3 min, but no longer than 12 h;
- Very long—last longer than 12 h, but no longer than 24 h; and
- Catastrophic—last longer than 24 h.

In addition to the mentioned indicators, other quantities are recorded, such as current harmonics, THDI current distortion factor, powers (active, reactive and apparent), PF power factor [10–13]. These quantities are given for informational purposes only and do not belong to the set of criterion quantities used to assess the quality of electricity supply. The requirements of [9] are taken as the basis for evaluating power supply conditions.

4. Regulatory Conditions

The basic legal act applicable to Polish law is the Regulation of the Minister of Climate and Environment [14]. In Poland, as in other EU countries, regulations in line with EU directives are applied. The EMC Directive (2014/30/EU) requires manufacturers, importers, and distributors to ensure that their products comply with EMC requirements. Products must be designed and manufactured in a way that ensures that they do not cause unacceptable electromagnetic interference and that they are immune to external interference. The manufacturer must draw up a declaration of conformity and label the product with the CE mark, which confirms compliance with the requirements of the EMC Directive. The equipment must undergo conformity assessment procedures, which may include internal production controls, laboratory testing, and certification by notified bodies.

In the case of equipment operating in underground mines, the assessment of hazards, the level of impact of this equipment on other receivers, and its resistance to interference should be particularly controlled. The environment of underground mines is special in terms of climatic and technical conditions and operational reliability, which directly affects the safety of working crews.

The requirements in Parts 2–4 of the standard [9] focus on compatibility levels for conducted interference in low-voltage power networks in industrial and public facilities. It defines the maximum permissible levels of electromagnetic interference that can occur in the power supply networks of these facilities. It covers interference, such as overvoltage conditions, harmonics, voltage rises, swings, and collapses, and other types of interference that can affect the operation of equipment. The standards describe the interference measurement methods and test procedures that must be carried out to assess compliance with the standard.

5. The Sources of Disturbance in a Mining Plant

Mining plants belong to that part of industrial plants where the configuration of the power grid changes continuously, depending on the progress of the mining process. This is due to the progress of mineral extraction, but also to the preparation of new areas and the decommissioning of old pits. These changes entail significant differences in the functioning of the networks and the impossibility of clearly predicting the phenomena that may occur in these networks.

The second major factor is the progressive mechanization and automation of work. This necessitates the use of advanced power electronic systems, which are a source of disturbances in mine networks, and also poses a threat to other power electronic systems that need power in the form of an undistorted sinusoid to function properly.

Machines that significantly affect the generation of disturbances in mine networks are hoisting machines and conveyor drives (Figure 2). In the first case, we are dealing with the cyclic switching on and off of a high-power drive. In the second case, it is a group of drives of significant power characterized by heavy starting, but without the ability to determine the moment of their stopping and restarting (this is due to the integration of these devices with other technological equipment or is caused by failures of these devices). The hoisting machine, as part of the mining vertical transportation system, is one of the largest consumers of electricity in a mining plant. Nowadays, in most cases, the drive motor is powered by a power electronic converter. It is a source of disturbances introduced into the plant network, and the consumer is exposed to disturbances occurring on the network. The literature does not describe much about the problems of disturbances in mining plant networks [15–17].

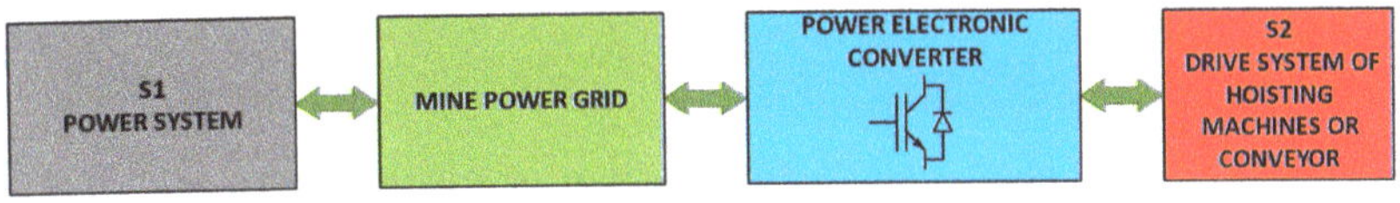

Figure 2. Connection of the energy source to the consumer through a power electronic circuit.

Figure 3 shows the waveforms of phase voltages recorded at one mining plant.

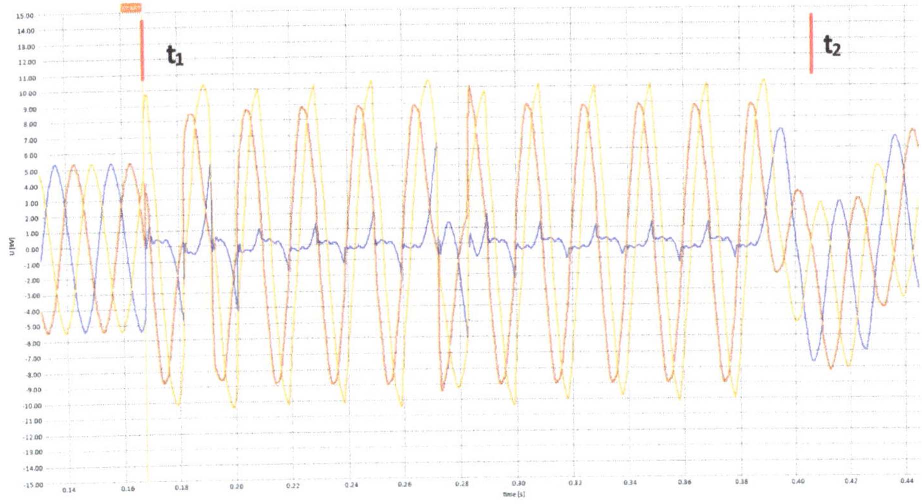

Figure 3. The waveforms of the three-phase voltages of the MV network.

Prior to the occurrence of the event (grid disturbance—at time t_1), the sinusoidal voltage waveforms will not cause any negative effects on the other consumers in the grid. However, during the time interval between t_1 and t_2, the operation of the power electronic converter will not be possible due to problems in synchronizing the operation of individual semiconductor valves of the converter. This means that subsequent components of the system will not be able to switch on properly, which may result in the interruption of the operation of the device and its emergency shutdown. In extreme cases, the semiconductor elements may be damaged. It is not difficult to imagine the consequences of such event if the converter mentioned before is a component of the hoisting machine drive system.

Deformations associated with changes in the shape of voltage waveforms are not the only ones that pose a threat to the correct operation of equipment. Another problem is the change in the effective value of the voltage (voltage rises, falls, and decays). The most common type of motors used in mines today are induction motors. The torque on the motor shaft depends on the square of the supply voltage. In addition, any change in the effective value of the supply voltage results in accelerated degradation of the insulation of such motors. These are hidden costs associated with the poor quality of the power supply.

Figures 4 and 5 show another dangerous phenomenon related to the quality problem, namely, voltage collapse, consisting of a temporary decrease or increase in the RMS value of the voltage. This can cause tripping of protections in network systems, and the frequent occurrence of such phenomena leads to a reduction in the life of equipment, including electric motors.

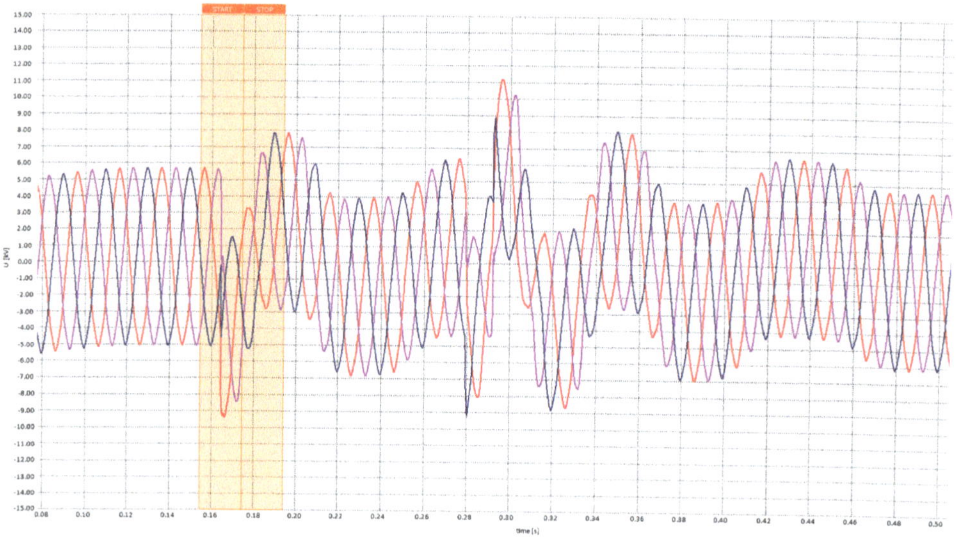

Figure 4. Voltage collapse (decrease in voltage value).

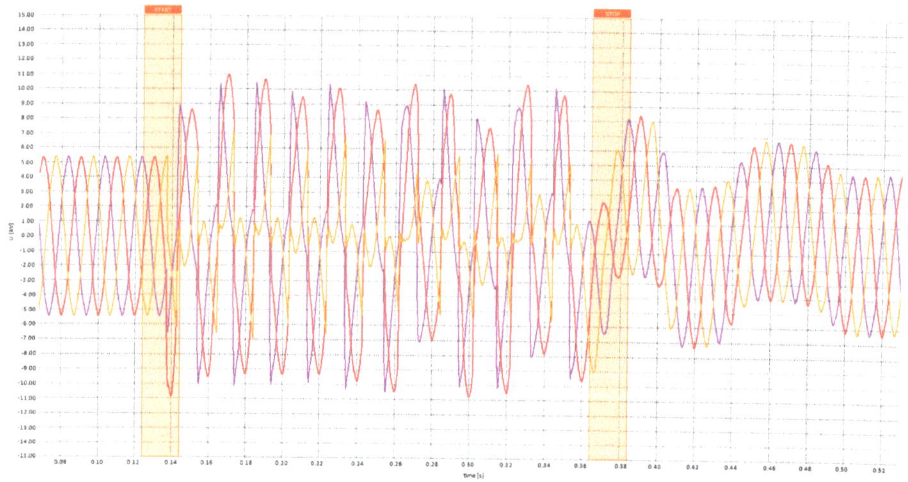

Figure 5. The waveform of phase voltages when the line voltage increases.

Another dangerous type of disturbance is a change in the frequency of the waveforms powering the equipment. Figure 6 shows the MV voltage waveforms. As a result of a previous disturbance in the network, there was a change in the frequency of the supply waveforms.

The frequency of supply voltage waveforms has a direct effect on the speed of induction motors. In mining plants, a great number of drives are directly connected to the grid, and the phenomenon shown in Figure 7 directly affects the instantaneous change in the speed of these machines.

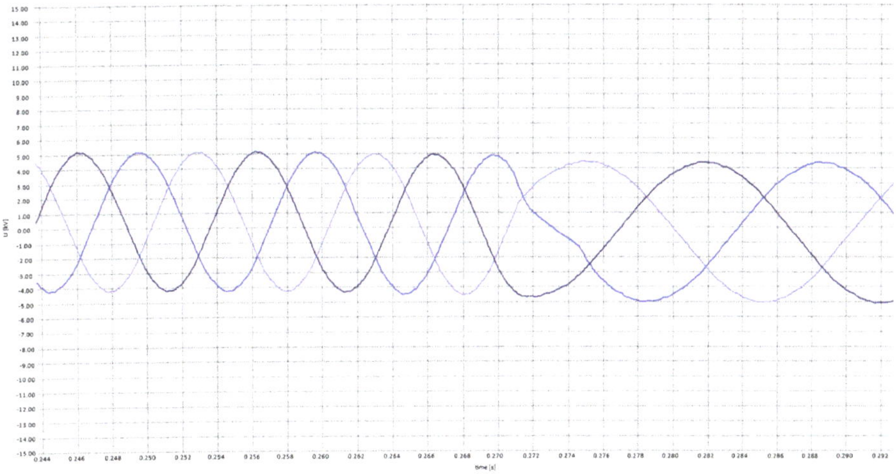

Figure 6. Change in the frequency of phase voltage waveforms in the MV network.

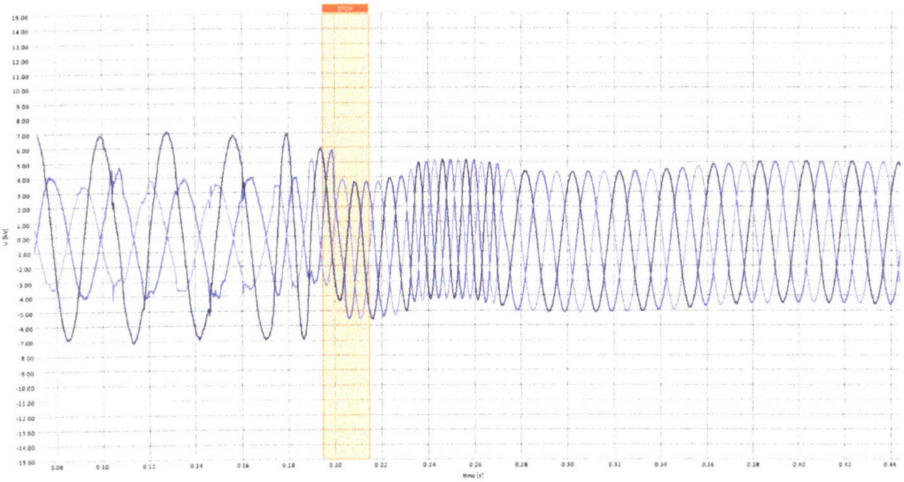

Figure 7. Disturbance (increase in network voltage).

The change in the frequency of the network waveforms shown in Figure 6 is the result of a sequence of events related to the increase in voltage, a significant deformation of the voltage waveforms, and a change in the frequency of the waveforms during the final phase of the entire event. Figure 6 shows the phase voltage waveforms during the entire event.

6. The Analysis of Disturbances at One of the Mining Plants

Taking into account the potential problems described earlier, during the discussion, possible courses of action were identified to minimize the potential adverse impacts of phenomena related to the quality of the electricity supply on the efficiency of the production process and the safety of the workforce.

In recent years, this mine has seen an increase in the number of power grid failures. At the same time, work was being carried out to expand the network in the underground workings. The analysis of the incidents in question showed that there may be problems in the networks of mining plants:

- problems with maintaining the appropriate effective voltage value,
- problems with insulation and its accelerated aging, and
- problems with deformation of voltage waveforms.

The consequences of the occurrence of these phenomena are significant economic losses due to uncontrolled interruptions of the production process. Sources of disturbances related to the quality of electricity supply in the plant include regulated electric drives of hoisting machines, belt conveyors, longwalls, and face complexes, as well as the extensive underground power system. In this case, hoisting machinery has been eliminated as a source of disturbance, due to the fact that these machines are built on the basis of the Leonard system.

The effect of a change in the effective value of the voltage mainly involves problems with the proper functioning of protection in mine networks and the reliable operation of contact elements of equipment. Due to the use of drives supplied directly from the mains, a change in the value of the supply voltage causes a change in the slip and an increase in current, which, as a consequence, causes increased heating of the machinery and a faster progressive wear process.

The largest number of failures, in this case, was caused by disturbances occurring in the underground power system. These primarily included damage to cables and wires, due to accelerated aging of the working insulation. Each failure triggered the tripping of protection in the switchgears and shut down the power supply to the mining region, stopping the operation of the entire process line as a result, and creating an additional hazard (shutting down the ventilation and drainage system). Re-energizing the power supply had to be preceded by many hours of preparation, including the localization of the damage site, the repair of the wire or cable by making a cable socket, and measurements of the network.

Another problem that potentially occurs in the networks of mining plants is the distortion of voltage waveforms. They can cause malfunctions of protections in power grids, problems with synchronization of converters, switching off of contact elements, and accelerated aging of insulation due to the flow of additional ground currents. All the aforementioned problems negatively affect the production process of mining plants. In addition, they reduce the reliability and operational safety of the mining plant.

The following events, among others, occurred in the power grid of the analyzed mining plant:

Incident 1

Emergency shutdown of one of the bays of the 6 kV switchgear.

The location of the fault was a 6 kV cable line leading from the field of the 6 kV switchgear, and emergency shutdown of the field occurred due to tripping of the earth fault protection. The cause was damage to the working insulation in one of the main cables supplying the underground workings. Visual inspection did not reveal mechanical damage to the armor, and no damage to the insulation sheath or fatigue due to stress was observed. The failure was the result of internal damage to the cable caused by the effect of the power quality on the acceleration of the ageing process of the working insulation. The consequences of the failure included no power supply to the 6 kV transformer stations connected to the cable line from the 6 kV switchyard. The consequence of the failure was the lack of power supply as follows:

- ore haulage equipment from the coalface;
- conveyors $2 \times 2 \times 100$ kW;
- pumps in the area of the exploratory heading (90 kW), ramps (90 kW), and local drainage in the face;
- equipment for the longwall complex; and
- fan for ventilation of the coalface.

The duration of failure was 2 h 30 min, representing the time required to locate the failure site and repair the damaged cable.

Incident 2

The second major failure occurred as a result of a 6 kV blackout in the main switchgear. As a result, power was disconnected to the equipment in the plant. The power was then started again. After a few minutes, one of the electricians found smoke coming from the hoisting machine building of one of the shafts. The power supply to the 6 kV switchboard and the power supply to the winding machine building in question were switched off.

The National Fire Service was called to the incident. After the smoke had been cleared from the winding machine building, it was visually established that the source of the smoke was an overheated starting choke of the synchronous motor of the winding machine control generator. In addition, it was noted that the power switch for the synchronous motor was in the ON state despite a power failure. The power failure caused the 6 kV circuit breaker, which switches off the synchronous motor starting choke during start-up, to switch off. The lack of excitation voltage (electromechanical exciter off) caused the motor to fall out of synchrony. Prolonged operation at currents several times the rated currents caused the starting choke windings to overheat.

As part of the removal of the failed components of the power supply system of the hoisting machine of the mine shaft lift, the following items were replaced:

- the power switch in the power supply system of the synchronous motor driving the control generator,
- 6 kV cable in the synchronous motor power supply system, and
- the starting choke.

After analyzing the incident, it was decided to assess the power supply quality indicators, which was carried out by the staff of the Centre for Electricity Quality operating within the AGH University of Krakow. On the basis of these findings, a modification was made to the mine's electrical network by changing the power supply structure and connecting some equipment to a different medium-voltage switchgear, which resulted in an improvement in the power supply quality indicators and thus a reduction in the number of failures in the plant network.

7. Measurement of Power Quality Indicators

The remainder of this article will present selected test results after changes were made to the mine's power supply system. Due to the fact that the mine consists of two parts, and each of them is supplied by different transformer stations, the measurements were carried out in several stages.

The analysis was carried out in terms of the presence of disturbances in the mine's power grid, which can affect the operation of equipment connected to it and cause negative financial consequences in the form of contractual penalties (reactive power, higher harmonics, and voltage fluctuations). The following indicators were adopted to assess the quality of the supply voltage:

- RMS value of the supply voltage,
- supply voltage asymmetry,
- supply voltage fluctuation,
- supply voltage distortion,
- supply voltage collapses and spikes, and
- power supply interruptions.

In addition to the mentioned indicators, the report also includes registrations of other quantities, such as supply voltage frequency, current harmonics, THDI current distortion factor, powers (active, reactive, and apparent), and power factor (PF). These quantities are given for informational purposes only and do not belong to the set of criteria used to assess power quality.

The basic time for aggregation of measurement data, based on the assessment of the correctness of power supply conditions that was carried out, was set at 10 min. Measurements were made with an aggregation of 1 min, so as to obtain more accurate information about the variability of the loads at the considered measurement points in case of a potential need to select compensating devices.

SONEL PQM-703 recorders (SONEL, Świdnica, Poland) were used for the measurements. The PQM-703 power quality recorder is a device for measuring, analyzing, and recording power network parameters and power quality in accordance with the standard [14]. The recorder fully complies with the requirements of [9] Class A.

The measurands, measuring ranges, and measurement accuracy of the analyzer are contained in [18].

Power quality analyzers were connected for simultaneous recording at four points of the mine's primary power supply system (Figure 8).

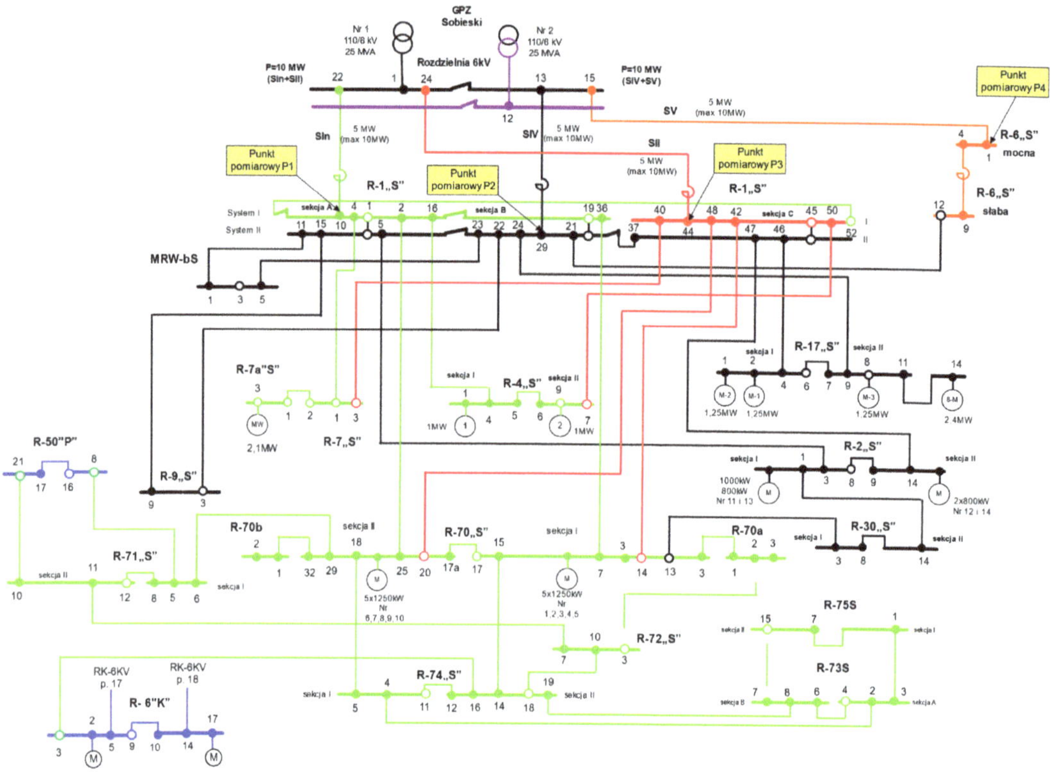

Figure 8. The installation sites of recorders used for measurements.

At each of the four points, one recorder was installed at the same time, and its current and voltage inputs were connected as shown in Figure 9.

Table 1 summarizes the results of each indicator for the four measurement points.

As can be seen in Table 1, the presented values of the individual coefficients are within the range of acceptable values, except for the coefficient of voltage asymmetry for point P4.

However, when analyzing networks with much smaller disturbances, it is clear that undesirable phenomena can occur in this network. This is due to the fact that the relevant coefficients take values close to the limit values.

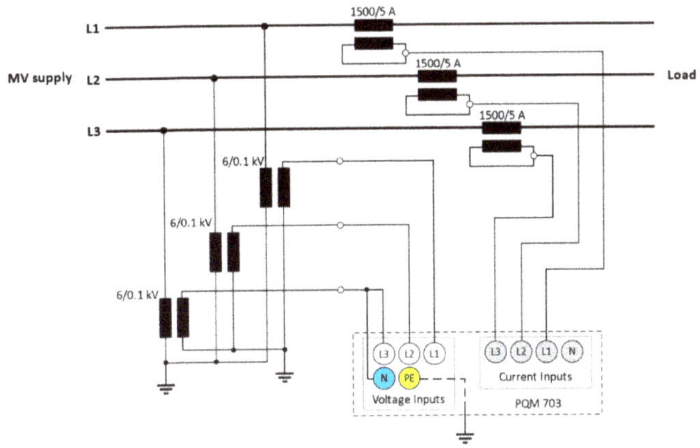

Figure 9. Diagram of the connection of the recorder at a selected point in the network.

Table 1. Summary of the numerical measurements (maximum values among the 3 phases) for which the assessment of compliance of the parameters of the quality of electricity supply is carried out.

Parameter		Measurement Point				The Permissible Level According to [9]
		P1	P2	P3	P4	
RMS value	CP0	6.184 kV	6.072 kV	6.029 kV	6.021 kV	5.1 kV
	CP100	6.399 kV	6.497 kV	6.506 kV	6.539 kV	6.6 kV
Voltage asymmetry	CP100	0.08%	0.07%	0.14%	3.7%	3%
Voltage fluctuations	CP95	0.84	0.84	0.87	0.92	1
Voltage distortion factor	CP100	0.95%	2.62%	3.42%	3.59%	10%

How the change in power supply structure affects the generation and propagation of disturbances in the grid can be seen in the waveform in Figure 10. Shown is the THDU waveform of one phase measured at measurement point P4. As can be seen, the extent of THD changes has clearly changed since 25 September, when the power supply structure was changed. It also depends on the current load. THDU changes between Monday and Friday are greater than those on Saturday and Sunday, which has to do with the supply of process equipment.

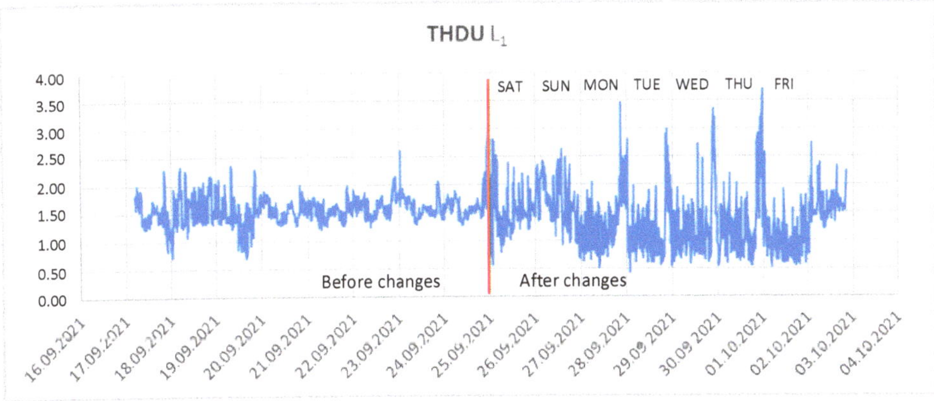

Figure 10. THD coefficient on the medium voltage side at point P4.

8. Conclusions

Undoubtedly, a problem that is very often raised during discussions on the quality of the electricity supply for mining plants is the issue of regulations that seek to regulate this problem in Poland holistically. Due to the development of renewable energy sources, regulations for the operation of the electricity system in Poland are being modified. However, this applies to the power supply of industrial plants, and the assessment of the problem of the system's impact on the power supply network is still not clearly defined. There are also standards available [9] that define the whole issue of power quality.

As can be seen from the analysis carried out, the normative conditions [9,14] may not be sufficient to ensure the safe and fault-free operation of equipment in underground mines. The conditions under which equipment operates vary greatly in terms of network extensiveness, power, and environmental conditions. An important aspect noted in this type of plant is the constant change in the configuration of the power network due to the progress of the work. In any case, all quality indicators and their changes over time should be analyzed. This will allow early detection of anomalies in the power system caused, for example, by changes to newer equipment or the installation of additional equipment. This is crucial in terms of ensuring the safety of the people working in these plants. In many cases, power failures and outages are not directly linked to power quality problems and are attributed to equipment failures.

In each case, failures result in a halt in mining, which translates directly into losses from unextracted ore. In the cases analyzed, problems resulting simply from poor power quality caused losses of more than PLN 1.5 million during the year. It is not possible to estimate the losses that could have occurred had appropriate measures not been taken to improve the voltage levels in the plant network, which had a direct impact on the safe and uninterrupted operation of the mine.

The design and commissioning of equipment for mining plants is based on regulations related to those plants. However, references to power quality issues cannot be found. Current mining plants will have to operate for another dozen or more years in a new reality, in which professional power engineering will force attention to matters regarding the quality of the power supply. The development of good models and, above all, the training of engineers by power and mechanical engineering departments to deal with such problems is no longer a luxury that only prosperous mining plants can afford, but is now becoming a necessity [19–22]. In the described mining plant, the changes made to the network are primarily authored by engineers associated with maintenance, and the necessary knowledge and measurement tools were provided to them by the AGH staff. The solution to the failure problem was the result of mutual cooperation and an exchange of experiences.

The currently used power quality indicators, or more precisely, their limits in the regulations, are not sufficient to ensure failure-free operation of mine power grids. To date, the costs of outages and failures have not been identified with the impact of poor power quality on power infrastructure. Ongoing studies show that the acceptable level of indicators is too high. The only problems, which have been misinterpreted as problems of poor power quality, have been excess levels of reactive power. This happened because of the penalties that distribution system operators Imposed on the mine. The paper presents a different approach to this issue and the need for careful analysis of measurement data, despite the initial compliance of the results with normative recommendations. Currently, research is being conducted on the possibility of using other indicators, such as aggregate indicators. However, this requires additional research and the development of guidelines for new regulations that will ensure an adequate level of power supply reliability under changing mine conditions.

Author Contributions: Conceptualization, T.S.; methodology, T.S.; software, J.W. and W.S.; validation, T.S., M.D. and J.W.; formal analysis, T.S., M.D. and J.W.; investigation, T.S., M.D. and J.W.; resources, T.S.; data curation, J.W. and W.S.; writing—original draft preparation, T.S., M.D., J.W. and W.S.; writing—review and editing, T.S.; visualization, J.W. and W.S.; supervision, T.S.; project administration, T.S.; funding acquisition, T.S. All authors have read and agreed to the published version of the manuscript.

Funding: This research received no external funding.

Data Availability Statement: The data presented in this study are available on request from the corresponding author. The data are not publicly available due to contractual restrictions on measurements.

Conflicts of Interest: The authors declare no conflicts of interest.

References

1. Kilter, J.; Meyer, J.; Elphick, S.; Milanovic, J.V. Guidelines for power quality monitoring—Results from CIGRE/CIRED JWG C4.112. In Proceedings of the 2014 16th International Conference on Harmonics and Quality of Power (ICHQP), Bucharest, Romania, 25–28 May 2014.
2. Martens, E.A., Jr.; Dias, L.F.S.; Fernandes, F.A.; Bonatto, B.D.; Abreu, J.P.G.; Arango, H. Evaluation and trends of power quality indices in distribution system. In Proceedings of the 9th International Coference. Electrical Power Quality and Utilisation, Barcelona, Spain, 9–11 October 2007.
3. Baggini, A. *Handbook of Power Quality*; John Wiley&Sons, Ltd.: Hoboken, NJ, USA, 2008; pp. 79–131, 187–259.
4. Dugan, R.C.; McGranaghan, M.F.; Santoso, S.; Wayne Beaty, H. *Electrical Power Systems Quality*; Mc-Graw-Hill: New York, NY, USA, 2004.
5. Li, H.; Lv, C.; Zhang, Y. Research on new characteristics of power quality in distribution network. In Proceedings of the IEEE International Conference on Power, Intelligent Computing and Systems (ICPICS), Shenyang, China, 12–14 July 2019.
6. Shepherd, W.; Zakikhani, P. *Energy Flow and Power Factor in Non-Sinusoidal Circuits*; Cambridge University Press: New York, NY, USA, 1979.
7. Hanzelka, Z. *Jakość Dostawy Energii Elektrycznej. Zaburzenia Wartości Skutecznej Napięcia*; Wydawnictwa AGH: Kraków, Poland, 2013.
8. Zhang, X.; Yan, Z. Energy quality: A definitione. *IEEE Open Access J. Power Energy* **2020**, *7*, 430–440. [CrossRef]
9. *IEC 61000-4-30*; Electromagnetic compatibility (EMC)—Part 4-30: Testing and Measurement Techniques—Power Quality Measurement Methods. PKN: Warszawa, Poland, 2015.
10. Karve, S. Harmonics—Active harmonic conditioners. In *Leonardo Power Quality Application Guide—Part 3.3.3*; A guide material by Leonardo Power Quality Initiative; European Copper Institute: Brussels, Belgium, 2001.
11. Hanzelka, Z.; Bien, A. Harmonics—Interharmonics. In *Leonardo Power Quality Application Guide—Part 3.3.1*; A guide material by Leonardo Power Quality Initiative; European Copper Institute: Brussels, Belgium, 2004.
12. Elmore, W.A.; Kramer, C.A.; Zocholl, E. Effect of waveform distortion on protective relays. *IEEE Trans. Ind. Appl.* **1993**, *29*, 404–411. [CrossRef]
13. Xu, W.; Liu, X.; Liu, Y. An investigation on the validity of power-direction method for harmonic source determination. *IEEE Trans. Power Deliv.* **2003**, *18*, 214–219. [CrossRef]
14. Regulation of the Minister of Climate and Environment of 22 March 2023, on Detailed Conditions for the Operation of the Electricity System. Available online: https://isap.sejm.gov.pl/isap.nsf/download.xsp/WDU20230000819/O/D20230819.pdf (accessed on 28 August 2024).
15. Bebikhov, Y.V.; Egorov, A.N.; Semenov, A.S. How higher harmonics affect the electrical facilities in mining power systems. In Proceedings of the International Conference on Industrial Engineering, Applications and Manufacturing (ICIEAM), Sochi, Russia, 18–22 May 2020.
16. Semenov, A.S.; Semenova, A.N.; Fedorov, O.V. The results of the implementation of the system for monitoring the quality of electricity in mining enterprises. In Proceedings of the 1st International Conference on Control Systems, Mathematical Modelling, Automation and Energy Efficiency (SUMMA), Lipetsk, Russia, 20–22 November 2019.
17. Wiechmann, E.P.; Aqueveque, P.; Munoz, L.; Henriquez, J.A. Energy quality and efficiency of an open pit mine distribution system: An improvement. In Proceedings of the IEEE Industry Application Society Annual Meeting, Vancouver, BC, Canada, 5–9 October 2014.
18. Instrukcja Obsługi. Analizatory Jakości Zasilania. Available online: https://www.sonel.pl (accessed on 28 August 2024).
19. Jiangzi, M.L. Measure and suppression of harmonics currents for heavy-duty hoisting machines. In Proceedings of the IEEE/PES Transmission and Distribution Conference & Exhibition: Asia and Pacific, Dalian, China, 26–29 November 2005.
20. Sukanth, T.; Jayanthu, S.; Jayalaxmi, A. Mitigation of power quality problem in underground mine using defferent control strategies. In Proceedings of the IEEE Region 10 Humanitarian Technology Conference (R10-HTC), Agra, India, 21–23 December 2016.

21. Siostrzonek, T.; Chmielowiec, K.; Firlit, A.; Barczentewicz, S. Zalety i wady stosowania układów wielopulsowych w napędzie elektrycznym maszyn wyciągowych. *Przegląd Elektrotechniczny Zeszyt* **2020**, *5*, 43–46. [CrossRef]
22. Siostrzonek, T.; Chmielowiec, K.; Piątek, K.; Dutka, M.; Firlit, A. The use of multi-pulse systems in the power supply of hoisting machine drives to improve voltage parameters in mining plants. In Proceedings of the EPQU'20, 12th International Conference and Exhibition on Electrical Power Quality and Utilization, Krakow, Poland, 14–15 September 2020.

energies

Article

Energy Consumption and Fume Analysis: A Comparative Analysis of the Blasting Technique and Mechanical Excavation in a Polish Gypsum Open-Pit Mine

Andrzej Biessikirski *, Przemysław Bodziony and Michał Dworzak

Faculty of Civil Engineering and Resource Management, AGH University of Krakow, 30-059 Krakow, Poland; przembo@agh.edu.pl (P.B.); dworzak@agh.edu.pl (M.D.)
* Correspondence: abiess@agh.edu.pl; Tel.: +48-12-617-20-70

Abstract: This article presents a comparative assessment of energy consumption and fume emissions such as NOx, CO_2, and CO associated with the excavation of a specified gypsum volume using two mining methods (blasting and mechanical extraction). The analysis was carried out based on a case study gypsum open-pit mine in Poland where both extraction methods are applied. The findings indicate that, for the same output volume, blasting operations require significantly less energy (ranging from 1298.12 MJ to 1462.22 MJ) compared to mechanical excavation (86,654.15 MJ). Furthermore, a substantial portion of the energy in blasting operations is attributed to explosive loading and drilling (970.95 MJ). Conversely, mechanical mining results in higher fume emissions compared to blasting. However, during mechanical extraction, the fumes are dispersed over a prolonged period of 275 h, whereas blasting achieves the same gypsum volume extraction in approximately 7.5 h. The prediction model suggests that, based on the obtained data, overall gypsum extraction will decline unless new operational levels are developed or the mine is expanded. This reduction in gypsum extraction will be accompanied by a corresponding decrease in energy consumption and emission of fumes.

Keywords: energy consumption; drilling; blasting; mechanical excavation; fumes

1. Introduction

Mineral extraction in open-pit mining is mostly related to the application of the blasting technique and mechanical extraction; however, drilling and blasting are considered the primary methods for quarrying rock mass due to their cost-effectiveness and ability to produce large volumes of well-fragmented rock in a relatively short period. This method is predominantly used for mining deposits of rock raw materials with compressive strengths above 20 MPa [1]. However, during the detonation process, explosive charges in boreholes generate a significant amount of energy, and, through the impact of the shock wave, stress waves, and expansion of the gaseous reaction products on the rock mass, they fracture the rock mass and release energy into the environment [2]. The literature shows that only 20–30% of the detonation energy is used for fragmentation, with the remaining energy being dispersed into the environment in the form of blast-induced vibrations, airborne shockwaves, and fly rock [3–6]. In sites where housing structures are close to open-pit mines, these environmental impacts can significantly restrict or even prohibit the use of blasting techniques [7–12]. Aside from the environmental risks associated with the release of detonation energy into the environment, the main gaseous products of decomposition reactions are large quantities of carbon and nitrogen oxides, which are considered fumes [13,14].

Where restrictions exist due to proximity to, e.g., housing structures or other constructions, quarry mining operations employ mechanical equipment, such as crawler tractors or bulldozers with ripper arms, excavators equipped with hydraulic breakers and vibration ripper devices, or continuous surface miners [15]. Compared to drilling and blasting, mechanical excavation is considered to be more expensive and time-consuming, with its main environmental impact associated with the production of fumes from diesel engines or near-distance emission of noise and dust [16–18]. The production rate of mechanical extraction in quarries depends on the type of machine used, its size, the mining and geological conditions, and the skills of the operators, based on which it varies from 10 to 500 $Mg \cdot h^{-1}$ [19–21]. The most common heavy equipment used in quarries is powered by diesel combustion engines, which are estimated to be 35–45% energy efficient [22–25].

Energy consumption in mechanical extraction by applying dozers with ripper attachments and/or eccentric vibratory skidders is widely discussed in the literature [21,26–30]. In general, ripping with crawler dozers is used to loosen the rock mass with a ripper and then move the loosened rock mass for loading, using single-head excavators or loaders, onto process haulers. Susceptibility to ripping is influenced by the physical properties of the rock, i.e., compressive strength, bedding plane, fracturing and cracking, brittleness and crystalline content, hardness, and weathering susceptibility. These parameters have a significant influence on the selection of the cavity size and the fracture angle of the ripper [21,26,31]. In the process of ripping, the ripper blade acts on a certain space of the rock medium in the space in front of the ripper and from its sides. As a result, the structure of the medium is destroyed to a degree that depends on the type of rock being ripped. In sum, these interactions constitute the magnitude of the ripping resistance [27]. The greatest efficiency of excavation with a crawler dozer with a skidder attachment is obtained by excavating the rock center in the direction of the fall of the deposited layers. However, the application of this method is associated with a high risk of repair downtime due to frequent damage and the accelerated wear of skidder attachment components. On the other hand, a vibratory ripper is an attachment fastened to the boom of a hydraulic backhoe excavator that is used to rip rock using its natural fractures. The process of ripping is based on the compression of the wedge of the working tool by the eccentric rotation of a shaft driven by a hydraulic motor. The process of destruction of the rock structure by the wedge tool is a dynamic phenomenon, and it is difficult to clearly determine the impact force and, therefore, the individual components of the resistance acting on the skidder [28–30]. Its value is determined by the unit penetration resistance—the resistance related to the unit length of the blade and the unit size of the penetration into the rock medium. This indicator depends mainly on the compressive strength of the rock and the natural cracks present in calcite. In order to determine the approximate mining efficiency of an eccentric vibratory skidder, it is necessary to determine the impact energy of the mining tool and the maximum force acting on the tool at the final stage of its penetration into the rock to a depth which depends on the type of equipment [27].

Comparative studies on mechanical mining versus blasting works have mainly been carried out for underground hard-rock mining. Stewart et al. compared the progress of excavation by TBM and blasting and highlighted the relevance of hybrid methods that combine both excavation technologies based on tunneling [32]. A similar comparison was made by Wennmohs [33] and Feijóo [34]. Skawina et al. compared the tunnel progress of mechanical excavation by a modular mobile mining machine and blasting based on discrete event simulation [35]. An LCA analysis and comparison of the two quarrying methods for quarries was carried out by Bascompta et al. and identified the critical factors in the emissions caused by blasting and mechanical excavation [36]. However, no studies have been carried out comparing the energy consumption of drilling and blasting and mechanical mining in quarry rock extraction operations.

This research paper presents a comparative analysis of the energy consumption of two mining methods—excavation with blasting techniques and mechanical extraction—based on the case study of an open-pit gypsum mine located in Poland. The energy consumption for excavating a selected section of the deposit with different mining methods was compared based on real data. Furthermore, the environmental impact of excavation by the selected methods was estimated based on the amount of gaseous products released into the environment in the form of carbon and nitrogen oxides. The energy consumption and the fumes were selected for comparison due to their applicability in both mining methods and the vulnerability of other performance factors to fluctuations in mining, geological, and market conditions and electricity and gas prices. This study aims to determine and compare the energy consumption and environmental impact due to fume emissions from mechanical excavation, drilling, and blasting, based on the case study of a Polish gypsum surface mine. The findings on energy consumption for both mining methods are of particular relevance to open-pit mine authorities, given the increasing costs of blasting per ton of output. Furthermore, the predictive model for overall output, based on data from the past five years, provides valuable insights into potential opportunities and risks at the operational level of the mine.

2. Materials and Methods

2.1. Test Stand

The evaluation of energy consumption was conducted for a gypsum open-pit mine located in the southern part of Poland. The proximity of the mine to residential structures imposes restrictions on the use of blasting operations. Consequently, extraction is performed using a combination of mechanical methods and blasting techniques. The area where these combined extraction methods are applied is depicted in red in Figure 1.

Figure 1. Open-pit mine's aerial view.

The gypsum deposit exhibits a gypsum content ranging from 75% to 99%, as illustrated in Figure 2 The uppermost layer of the deposit is primarily composed of low-calcination gypsum with a purity between 75% and 85%. Additionally, this layer is characterized by the presence of various geological features such as intrusions and sinkholes. The quality of the deposit improves progressively with depth (from the second operational level), reaching gypsum purity levels between 90% and 99%.

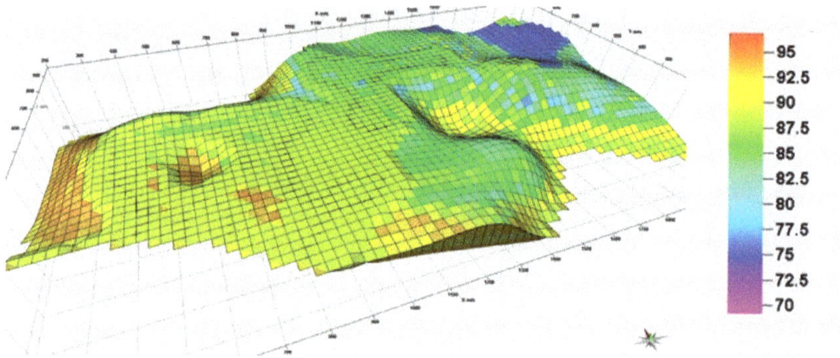

Figure 2. An overview of the deposit quality model.

The jaw crusher utilized in the open-pit mining operation specifies that all quarried material with a diameter of approximately 650 mm or greater is classified as overburden material.

2.1.1. Blasting Works

Blasting operations, shown in Figure 3, were carried out by a drilling and blasting company active on the Polish market. Single-row, long-borehole blasting was employed. Explosive charges were undivided, with each series comprising 11 boreholes loaded with ANFO (ammonium nitrate fuel oil). The characteristics of the ANFO components are detailed in Section 2.2. Each ANFO charge was primed using an electronic detonator and a 0.5 kg TNT booster. The blasting pattern parameters, listed in Table 1, were provided directly by the mine authorities and represent the actual parameters employed on-site.

Figure 3. Blasting works at open-pit mine.

Given the proximity of the mine to the residential structures (top side of Figure 1) and the road (right side of Figure 1), the adoption of mechanical mining methods has become increasingly prevalent.

Table 1. Blasting pattern.

Name of Parameter	Value
Bench height, m	8.0
Borehole diameter, mm	95.0
Length of borehole with subdrill, m	9.0
Stemming length, m	1.0
Burden, m	3.5
Spacing, m	3.8
Mass of explosive per borehole, kg	37.0
Mass of explosive per delay, kg	37.0
Total mass of explosive, kg	407.0
Number of boreholes in series	11
Number of rows	1
Gypsum volume, tons	12,779

2.1.2. Mechanical Extraction

In a research case study, the widespread combination of a crawler dozer equipped with ripper attachments and an excavator equipped with a hydraulic eccentric skidder was used. The technological machinery system extracted an average of 12,779 tons of gypsum over a calendar year. This extraction was achieved over a cumulative operational period of approximately 1350 mths (hours of operation). The information about mechanical extraction is provided in Table 2.

Table 2. Mechanical extraction.

Name of Parameter	Value
Annual gypsum extraction, tons	12,779
Bench height, m	8.0
Burden, m	3.5
Bench length, m	41.8
Gypsum volume, tons	12,779

Detailed data on fuel consumption (omitting the fuel consumption for other technological processes) and the quantity of gypsum extracted by each mechanical system are presented in Table 3.

Table 3. Summary of machine output for annual gypsum extraction and corresponding fuel consumption for each machine.

Parameter	Caterpillar 330DL	Caterpillar D9R
Annual volume of gypsum extraction, Mg	6711	6068
Value of fuel consumption, l	32,253	26,362

The mechanical equipment, which was applied in the blasting and mechanical operations, shown in Figure 4, is described in detail in Section 2.3.

Figure 4. Mechanical extraction at the open-pit mine.

2.2. Materials

Ammonium nitrate (V) was produced by Yara International ASA (Szczecin, Poland). The sample contained approximately 35.0% nitrogen, with a prill diameter of 1 mm and a bulk density of 820 kg·m^{-3}.

The fuel oil (FO) was delivered by one of the operators in the Polish market. This FO sample consisted of a wide-range fraction of C10-C20 hydrocarbons, with a bulk density of 800 kg·m^{-3}. The kinetic viscosity of the FO sample was measured at 13.6 mm^2·s^{-1} at 40 °C.

ANFO was blended in the Universal Mixing System (UMS). The system allows the manufacturing of ANFO or emulsion bulk explosive on-site. Fuel oil was blended with ammonium nitrate (V) straight before the loading process. The blending ratio of ammonium nitrate (V) to FO was 94.0:6.0 (% wt.).

2.3. Equipment

The Hausherr HSB 111P was used in the drilling operations. The HSB 111P is a crawler-mounted, hydraulic top-head drive drilling rig. The drilling rig had a gross power output of 125 kW. The average diesel consumption of the rig was approximately 12 l·h^{-1}. The engine of the drilling rig complied with the Tier 1/Final Stage I emission standards. The technical specifications of the drilling rig are detailed in Table 4.

Table 4. Summary of the technical parameters of the machines included in the analysis.

Technical Parameters	Caterpillar 330DL	Caterpillar D9R	Hausherr HSB111-P
Gross power, kW	200	302	125
Average diesel consumption, l·h^{-1}	22.76	26.76	12.32
Total efficiency of the powertrain unit η, %	35	37	38

The Mobile Explosive Mixing Unit (MEMU) is a truck designed for the on-site production and loading of ANFO, emulsion bulk, and heavy ANFO explosives. This unit was mounted on the Scania chassis, which provided a gross power output of 353 kW. The efficiency of the MEMU in manufacturing and loading ANFO averaged between 70 and 80 kg·h^{-1}. During loading and blending operations, the fuel consumption of the MEMU was typically between 30 and 35 l·h^{-1}.

The Caterpillar D9R crawler dozer is a versatile machine engineered for a range of mining operations including ripping overburden or production dozing. The dozer complies with Tier 3/Final Stage IIIA emissions requirements. It boasts a gross power output of 337 kW and has an operating weight of approximately 49.99 tons. Fuel consumption according to technical data varies depending on the type of operation (light, medium, heavy), ranging from 32 to 46 l h^{-1}; however, the average fuel consumption obtained from data from the mine site was determined to be 25 l h^{-1}. The technical specifications of the crawler dozer, which was equipped with a ripper, are detailed in Table 4.

The Caterpillar 330D is a hydraulic excavator equipped with an XC 30 excentric ripper. This excavator has an approximate operating weight of 35.3 tons and a gross power output of 200 kW. The average fuel consumption is approximately 23 l·h^{-1}. The Caterpillar 330D complies with the Tier 3/Final Stage IIIA emission standards. The technical specifications are detailed in Table 4.

2.4. Methods

2.4.1. Energy Consumption and Fumes

Fumes (CO_x and NO_x) were quantified in accordance with the standard [37], which adheres to directive [38]. ANFO charges, each weighing 500 g, were detonated within a steel mortar situated in the blasting chamber. Following detonation, the resulting fumes were homogenized for 3 min using a mixing procedure. Subsequently, gasses were collected for 20 min via the ventilation system. The concentrations of CO_x and NOx were determined using IR (MIR 25e, ENVEA, Paris, France) and chemiluminescent (TOPAZE 32M, ENVEA, Paris, France) analyzers, respectively. The concentrations of carbon monoxide (CO) and nitrogen oxides (NO_x) in the gaseous products were measured and reported per kilogram of explosive used.

The explosion energy was evaluated using a detonation calorimeter.

The method enables the assessment of the detonation energy generated from the detonation of 1 kg of explosive material. A 50 g sample of non-ideal explosive was initiated with an electric detonator and a 5 g RDX (Royal Detonation Explosive) booster. The results were recorded by considering the characteristic temperatures of four cycles (T_1–T_4), the duration of the main period of the cycle (n), and the total heat effect (Q) calculated for the predetermined heat capacity of the assembly (K). These parameters were noted upon the completion of the measurement. The total heat effect was calculated using Equation (2).

$$Q = K \cdot (\Delta T - k) \tag{1}$$

In the above equation, K is the heat capacity of the assembly, in cal °C; ΔT is the variation in temperature during the main period of the cycle, where $\Delta T = T_3 – T_2$, °C; T_1, T_2, T_3, and T_4 are the specific temperatures in characteristic points of the measurement cycle, in °C; and k is the coefficient which is responsible for making a correction on losses of the assembly, in °C, computed based on Equation (2):

$$k = 0.5 \cdot [0.2 \cdot (T_2 - T_1) + 0.2 \cdot (T_4 - T_2)] + (0.2 \cdot (n-1) \cdot (T_4 - T_3)) \tag{2}$$

In the above equation, n denotes the duration of the main period of the cycle, in min.

The total efficiency of the powertrain systems, as well as the efficiency of the hydraulic working systems of the machines, was assessed based on the technical and operational documentation provided by the manufacturers. The drivetrain systems of the analyzed machines were evaluated by considering their drivetrain structure as a mixed, series-parallel configuration, which was calculated according to Equation (3). In these systems, the drivetrain components, together with the motor, are responsible for the conversion and transmission of energy from the driving components to the passive components, which are subjected to resistance forces. However, not all the work performed by the active forces is used for the intended purposes. A portion of the energy is expended in overcoming frictional resistance during movement, which is dissipated as heat into the environment [39]:

$$\eta = \prod_{j=1}^{n} \eta ij \tag{3}$$

where η is the total efficiency of the powertrain unit in %; n is the number of powertrain components; and η_{ij} is a multiplication of the efficiency of individual components of the powertrain unit.

The analyzed machines were equipped with diesel engines meeting the Tier 3/Stage IIIA/B emission standard. Fume (NO_x and CO_x) emissions were determined based on Equation (4), used for the Tier 3 methodology [40]:

$$M_E = N \cdot HRS \cdot P \cdot (1 + DFA) \cdot LFA \cdot EFBase \tag{4}$$

In the above equation, the following definitions are used:

M_E is the mass of emissions of pollutant during inventory period in $g \cdot h^{-1}$;

N is the number of engine units;

HRS is the annual hours of use in h;

P is the engine size in kW;

DFA is the deterioration factor adjustment;

LFA is the load factor adjustment;

$EFBase$ is the vase emission factor in $g \cdot kWh^{-1}$.

The parameters N, HRS, P, DFA, LFA, and $EFBase$ are split further by classification systems as detailed below.

The machinery/vehicle population (N) is split into different technology levels and power ranges; the annual working hour (HRS) is a function of the age of the equipment/vehicles, so, for each subcategory, individual age-dependent usage patterns can be defined; power (P) is a function of the power distribution of the vehicles/machinery, so, for each subcategory, an individual power distribution can be defined within the given power ranges; the emission factor ($EFBase$) is determined for each pollutant as a function of technology levels and power output; finally, the deterioration factor adjustment (DFA) is a function of the power range of the vehicles/machinery and the technology level.

Table 5 and Figure 5 present the emissions of NO_x, CO, and CO_2 produced during one hour of effective operation in both mechanical and blasting processes.

Table 5. The volume of fume emissions related to the operating time of the analyzed machines.

Volume of GHG Emissions of Gases	Caterpillar 330DL	Caterpillar D9R	Hausherr HSB111-P
CO, $g \cdot h^{-1}$	300	450	125
CO_2, $g \cdot h^{-1}$	6140	7150	34,011
NO_x, $g \cdot h^{-1}$	648	544	38

2.4.2. Prediction Model Method

The prediction models were determined using the FBProphet algorithm (based on the FBProphet module in Python 3).

The FBProphet algorithm is an additive regression model featuring either a piecewise linear or logistic growth trend. This model incorporates a yearly seasonal component, represented via Fourier series, and a weekly seasonal component, modeled using dummy variables. Due to the limited dataset provided by the open-pit mine authorities, a full seasonal analysis was not included in the manuscript. Generally, the Prophet model is well-suited for datasets that encompass extended time periods (months or years), exhibit multiple strong seasonality, include known significant yet irregular events, contain missing data points or significant outliers, or show non-linear growth trends approaching a limiting value. The algorithm was employed for modeling time series data as a combination of trend, seasonality, and noise components, utilizing Bayesian inference. It applied a decomposable time series model described by Harvey and Peters [41,42], according to Equation (5) [43]:

$$y(t) = g(t) + e(t) + h(t) + E_t \tag{5}$$

where $y(t)$ represents the value of the time series at a specific time; $g(t)$ is the trend component; $e(t)$ is the error term; $s(t)$ indicates the periodic changes; $h(t)$ represents the effects with irregular schedules; and E_t is an error term that accounts for any unusual changes not accommodated within the model [43]. As previously mentioned, the season analysis, due to the size of the dataset, was omitted.

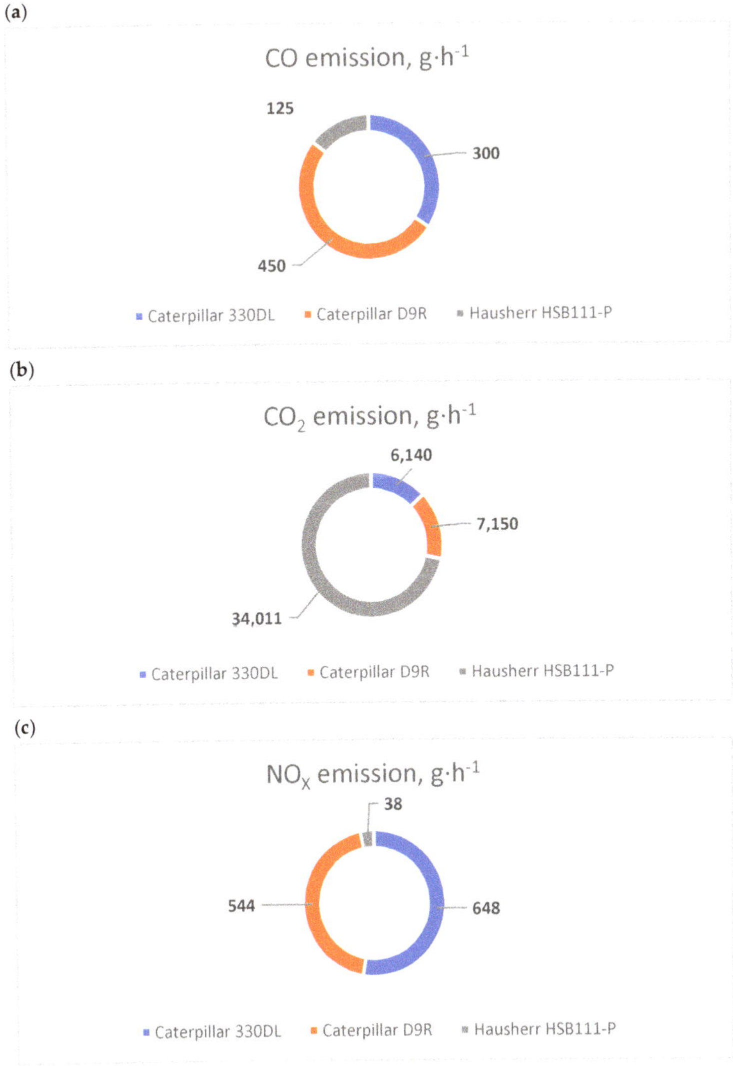

Figure 5. Fumes emitted from mechanical equipment during one hour of operation time: (**a**) CO, (**b**) CO_2, and (**c**) NO_x.

The trend was modeled using a piecewise linear regression approach, allowing it to be represented as a sequence of linear segments. The slope of each segment was a function of change points within the time series data. The mathematical representation of the trend follows Equation (6):

$$y(t) = g(t) + e(t) \tag{6}$$

The trend component was modeled as a piecewise linear function, described by Equation (7):

$$g(t) = (k(t) \cdot t) + m(t) \tag{7}$$

where $k(t)$ defines the slope of the trend at time (t), and $m(t)$ represents the intercept at time t. Both the slope and the intercept were modeled using a hierarchical Bayesian model. This approach allowed for the regularization of estimators and captured the uncertainty surrounding them.

The growth function in the prediction model modelled the overall data trend. By default, the algorithm applies a linear growth model, using piecewise linear equations with varying slopes at specific change points. When a time series approaches a saturation level, it means that the values are constrained by a maximum (cap) or minimum (floor), and a logistic growth function is preferable, as it accounts for these upper and lower limits. If no trend growth is observed, a flat growth function is applied, resulting in a constant trend value over time. For the current analysis, csv files containing either energy consumption or the annual gypsum extraction volume were provided to model these growth dynamics.

3. Results and Discussion

3.1. Blasting Techniques

Blasting operations are a crucial aspect of mining activities, playing a significant role in the efficiency of resource extraction and the overall energy consumption of the mining process. The energy consumption in blasting is multifaceted, encompassing the energy stored in the explosives, the energy required for drilling boreholes, and the energy expended in post-blast activities such as mucking and hauling. However, for this analysis, the energy related to post-blast activities will be omitted due to the similarity of these processes to those performed during mechanical extraction at the mine site.

3.1.1. Evaluation of Explosion Energy of Energetic Materials

The primary energy input in blasting operations is the energy stored in the applied explosives. The energy released from explosives is crucial for fragmenting the rock mass and facilitating subsequent extraction processes. The total explosive energy was calculated based on detonation calorimetry tests, providing a measure of the energy content of the explosives used. This approach enables the evaluation of the actual energy output from the tested blasting series.

Table 6 presents the specific energy release values for 1 kg of the applied explosives. Based on the detonation calorimetry measurements, the energy release for ANFO (ammonium nitrate fuel oil) is approximately 3940 kJ·kg^{-1}. In one blasting series, a total of 407 kg of ANFO was used, resulting in an energy release of 1603.58 MJ. In comparison, the TNT booster generated approximately 6400 kJ kg^{-1}. For the same blasting series, a total of 5.5 kg of TNT was used, producing an energy release of 35.2 MJ. Combining both explosives, the total mass of ANFO and TNT charges used was 412.5 kg, yielding an estimated stored energy in the explosives of 2635.6 MJ.

Table 6. Average energy of explosion.

Energetic Material	Average Energy of Explosion, kJ·kg^{-1}	Maximum Deflection, %
ANFO	3940	0.5
TNT *	6400	-

* TNT energy of explosive based on OZM research [44].

It is important to note that not all the stored energy in the explosives is utilized for quarrying. According to studies by Rai and Singh, Pyra, and Smujłło, approximately 70–80% of the energy is dissipated into the environment in the form of air blast, blast-induced vibrations, and fly rock. Therefore, the effective energy available for quarrying is only 20–30% of the total stored energy [5,45,46].

From the total energy value of 1638.78 MJ, the effective energy for quarrying, calculated as 20–30% of the total, ranges from 327.76 MJ to 491.63 MJ. This indicates that a significant portion of the energy is lost to non-productive forms, emphasizing the need for optimizing blasting operations to enhance energy efficiency.

3.1.2. Evaluation of Energy Consumption of Loading and Drilling Operations

The energy expenditure required to excavate part of the deposit is determined by Equation (8), taking into account the effective energy of detonation of the blasting net of explosives and the energy required to drill and load the blasting net of the holes, as well as the total efficiency of the drive systems of the machines used, including the fuel and total efficiency. Drilling boreholes is essential in overall blasting operations. Energy consumption in drilling depends on several factors, including the type of drilling equipment used, the geological characteristics of the rock mass, and the borehole diameter and depth. In general, energy consumption from blasting operations can be presented according to Equations (8) and (9):

$$E_{TB} = E_{Det} + E_{Drilling} + E_{Loading} \tag{8}$$

$$E_{TB} = \alpha \cdot E_{Ex} \cdot Q_T + \sum_i \left(\eta_i \cdot \underline{V_{FO_i}} \cdot d_{FOi} \cdot CV_{FOi} \right) \tag{9}$$

In the above equations, the following definitions are used:

E_{TB} is a total sum of the energy consumption from all blasting operations in MJ;

E_{Det} is the energy released from the detonation of explosives in MJ;

$E_{Drilling}$ is the energy consumption from drilling operation in MJ;

$E_{Loading}$ is the energy consumption from loading operations in MJ;

α is an index for quarrying (energy which is subjected only for extraction 0.2–0.3, according to [21,26,28]);

Q_T is the energy of the explosion in MJ·g^{-1};

η_i is the total efficiency of the powertrain unit in %;

$\underline{V_{FOi}}$ is the average volume of combusted diesel oil in dm^3·h^{-1};

d_{FOi} is the fuel oil density in kg·dm^3;

CV_{FOi} is the calorific value of combusted fuel oil in MJ·kg^{-1}.

According to Table 3, the Hausherr 111P drill rig has a drilling efficiency of 23.18 m/h and an average fuel consumption of 11.57 dm^3/mth drilled.

For an analyzed series requiring a total borehole length of 99 m (11 boreholes), the drilling operation (assuming no maneuvering time) takes approximately 256 min. This results in a total diesel consumption of 49.40 dm^3. Given the average density of diesel at 0.8325 kg·dm^3 and a calorific value of 42.6 MJ·kg^{-1}, the drilling process consumes 41.13 kg of diesel, corresponding to an energy requirement of 1752.14 MJ.

Considering that diesel engines and power transmission systems for Hausherr 111P have an efficiency of 38%, as shown in Table 4, the effective energy utilized for drilling is approximately 665.81 MJ.

The process of loading explosives into boreholes involves the handling and placement of explosive materials, such as ANFO and boosters, and is generally less energy-intensive compared to drilling. The Mobile Explosive Manufacturing Unit which is utilized at the blasting site in this case study operates for an average of 35 min (to produce and load 11 boreholes with ANFO). During this period, the mobile unit consumes an average of 32.5 dm^3 of diesel oil. Given the diesel oil's density of approximately 0.8325 MJ·kg^{-1} and its calorific value of 42.6 MJ·kg^{-1}, the total mass of diesel combustion is 18.85 kg. This combustion results in a total energy output of 803.01 MJ [47].

However, considering the efficiency of diesel engines, which is typically around 38%, the effective energy available for the loading process is approximately 305.14 MJ. This energy estimation underscores the relatively lower energy demands of the loading phase compared to the drilling phase in blasting operations and is close to the energy range of 327.76 MJ to 491.63 MJ stored in explosives.

Analyzing the overall energy consumption of the loading and drilling process, detailed in Table 7 and Figure 6, it can be stated that drilling and loading actions require around 970.95 MJ of energy. The calculated value of energy consumption which is derived from loading and drilling actions is 2.5–3 times greater than the energy released from the explosive material which is subjected only to quarrying. However, the main difference is the process duration. In terms of the explosive decomposition reaction, the energy release is abrupt, which results in rock fragmentation.

Table 7. Average energy consumption of each blasting operation.

Operation	Energy Consumption, MJ
Detonation process (quarrying)	327.76–481.63
Drilling process	665.81
Explosive loading process	305.14

(**a**)

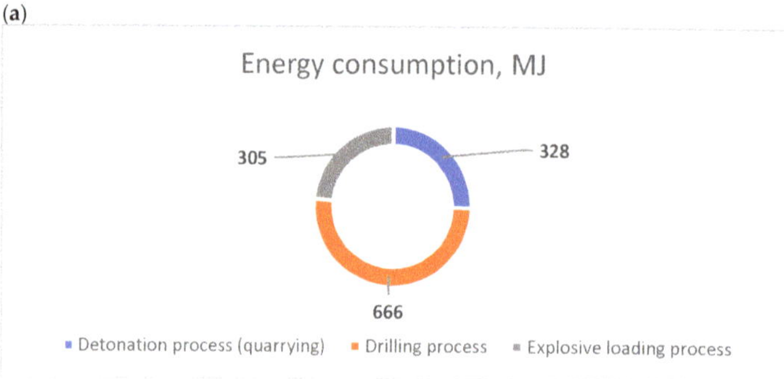

(**b**)

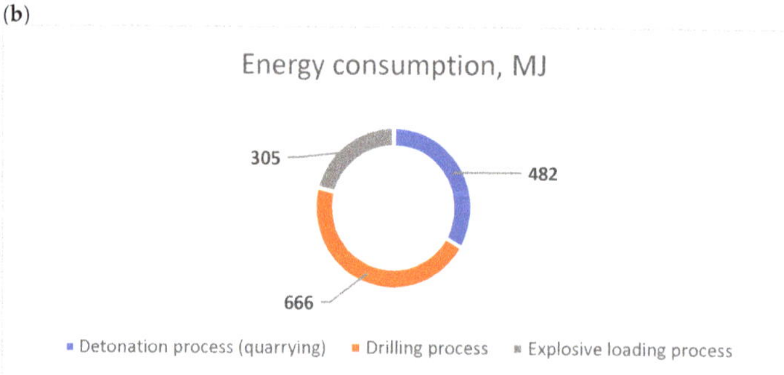

Figure 6. Total energy consumption of blasting works where the effective energy of detonation is (**a**) 20% and (**b**) 30%.

3.1.3. Evaluation of Energy Consumption of Loading and Drilling Operations

Blasting operations are a critical component of mining activities, serving as a fundamental method for breaking rock and facilitating the extraction of mineral resources. These operations encompass a series of processes, including drilling, explosive loading, and detonation, each contributing to the overall efficiency and effectiveness of the mining process. However, these operations also result in the generation of fumes and emissions, which pose significant health, safety, and environmental challenges.

The detonation of explosives in mining leads to the production of various fumes, the composition of which is influenced mainly by the type of explosive used and the efficiency of the detonation process. The composition of the fumes generated by the detonation of the applied explosives—ANFO (ammonium nitrate fuel oil) and TNT (trinitrotoluene)—is detailed in Table 8.

Table 8. Average fume volume formed during the decomposition reaction.

Energetic Material	Amount, $dm^3 \cdot kg^{-1}$		
	CO_2	CO	NO_x
ANFO	144.5	16.4	13.1
TNT	919.6	11.5	4.8

According to Table 7, the detonation of the explosives that were applied in series (ANFO and TNT) resulted in the release of various gaseous byproducts. The volumetric emissions per kilogram of ANFO were as follows: 144.5 dm^3 of carbon dioxide, 16.4 dm^3 of carbon monoxide, and 13.1 dm^3 of NO_x·, as detailed in Table 8 and Figure 7. In comparison to ANFO, 1 kg of TNT generated 919.6 dm^3 of CO_2, 11.5 dm^3 of CO, and 4.8 dm^3 of NO_x·. The observed disparity in carbon dioxide emissions between ANFO and TNT can be attributed to the oxygen balance inherent in these explosives. ANFO exhibits an approximately zero oxygen balance, leading to a more complete combustion process and, consequently, lower fume concentrations. Conversely, TNT possesses an oxygen balance of approximately −74%, indicating an oxygen-deficient decomposition reaction which results in a higher volume of carbon oxides (CO_x). Moreover, the chemical structure of the explosive significantly influences the composition and volume of the fumes generated. TNT's molecular configuration is rich in carbon atoms, accounting for its substantial CO_2 output. In contrast, ANFO's composition is nitrogen-rich, contributing to its distinct fume profile which includes higher nitrogen oxides. For a combined detonation of 412.5 kg of explosives, the estimated total fume volumes produced were 63.87 m^3 of CO_2, 6.74 m^3 of CO, and 5.36 m^3 of NO_x.

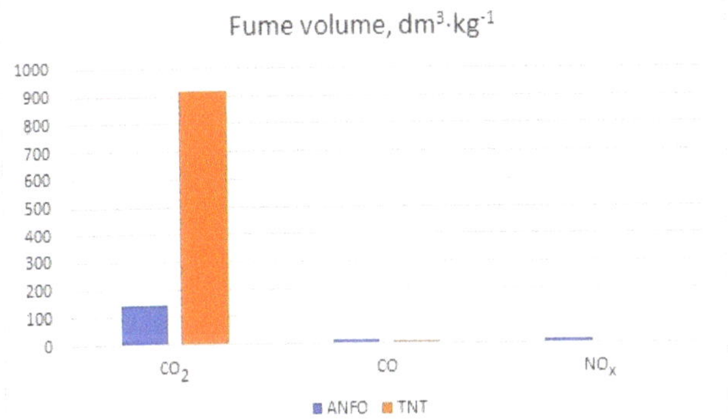

Figure 7. Fume volume comparison of ANFO and TNT fumes.

The emissions of NO_x and CO resulting from diesel combustion during drilling and loading operations were assessed based on parameters outlined in the literature, particularly referencing standards such as the Tier and Final stage. The assessment considered the engine horsepower and age for both the drilling rig and the Mobile Explosive Manufacturing Unit. In the case of the drilling rig, the NO_x emission was evaluated to be 50 $g \cdot h^{-1}$ and CO 187.5 $g \cdot h^{-1}$.The drilling of 11 boreholes required 256 min. This resulted in the emission of 0.21 kg of NO_x and 0.8 kg of CO. The combustion of diesel generated 2.67 kg of CO_2 per liter. For 41.13 kg of diesel fuel consumed, the CO_2 emission was 109.8 kg of CO_2.

In the case of loading operations, the NO_x emission was evaluated to be 60 g·h^{-1} and CO 225 g·h^{-1}. The loading process lasted for 35 min, resulting in the following emissions: 0.0348 kg of NO_x, 0.1305 kg of CO, and 50.3 kg of CO_2. Combining the emissions from drilling and loading operations with those from the detonation of explosives, the total emissions were calculated as follows: 10.48 kg for NO_x, 8.61 kg (density 1.91 g·dm^{-3} based on [48]) for CO, and 286.34 kg for CO_2, as detailed in Table 9 and Figure 8.

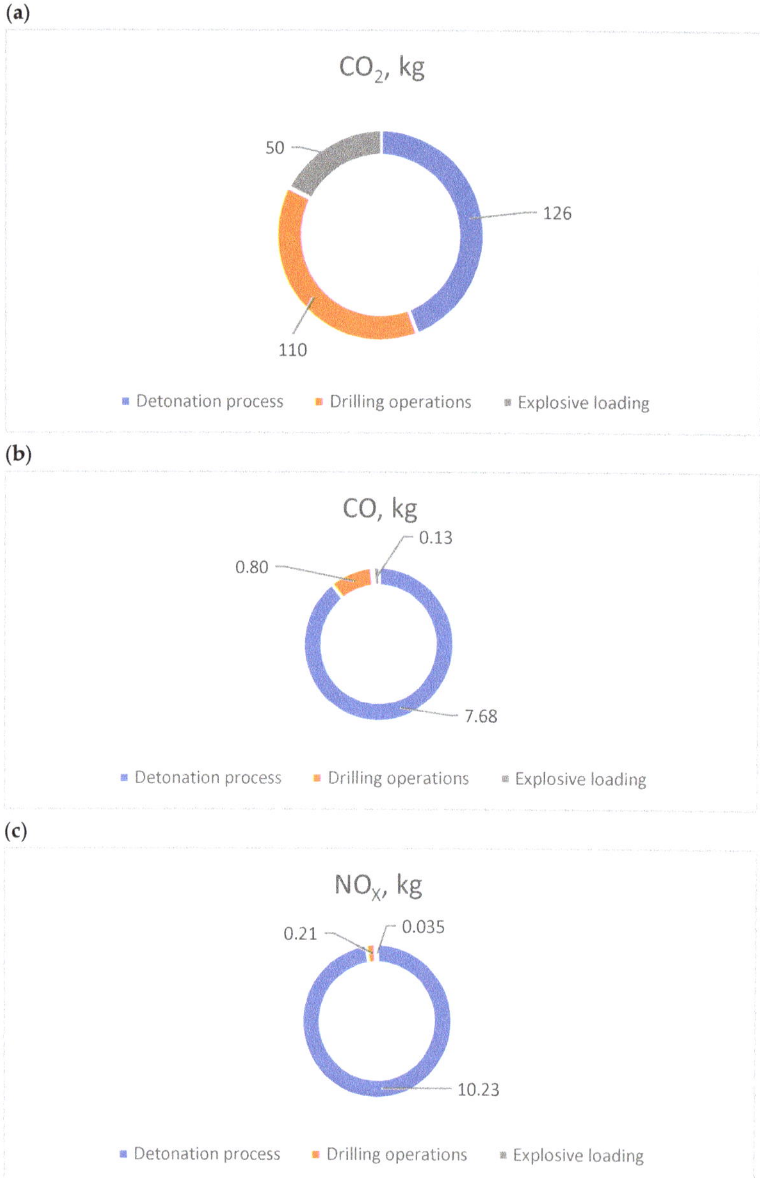

Figure 8. *Cont.*

(d)

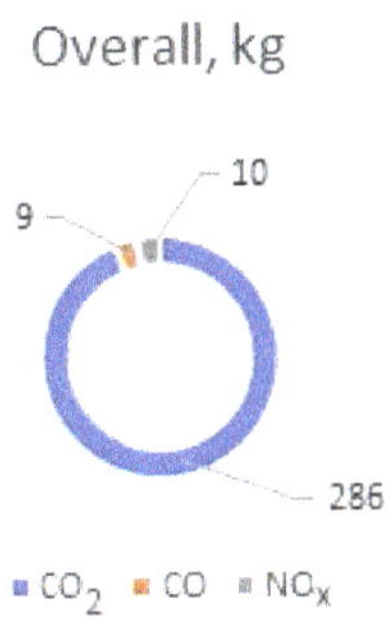

Figure 8. Total fumes emitted from all blasting operations: (**a**) CO, (**b**) CO_2, (**c**) NO_x, and (**d**) overall.

Table 9. Average fume emissions from this case study's blasting operations.

Type of Operation	Amount, kg		
	CO_2	CO	NO_x
Detonation process	126.24	7.68	10.23
Drilling operations	109.8	0.8	0.21
Explosive loading	50.3	0.130	0.035
Overall	286.34	8.61	10.48

This analysis revealed that, while drilling and loading operations contribute to fume emissions, the detonation process is the primary source of rapid and concentrated fume emissions, particularly impacting air quality in the vicinity of the blast site. Unlike the gradual emission and dispersion from drilling and loading, the sudden release of gasses during detonation poses a more immediate hazard to workers and necessitates stringent safety measures.

3.2. Mechanical Excavation

3.2.1. Evaluation of Mechanical Excavation's Energy Consumption

Mechanical excavation in open-pit mining involves the use of various types of heavy machinery and equipment, each contributing to the total energy consumption of the mining operation. The energy usage in such operations is primarily dependent on the types of machinery employed, the nature of the deposit, the scale of the mining activities, and the operational practices followed. In the specific case study examined, gypsum excavation was conducted through the coordinated operation of a crawler bulldozer and a crawler excavator. It is noteworthy that hauling operations were not considered within the scope of this analysis, since they remain a requisite of both mechanical and blasting excavation methods. In general, energy consumption from mechanical excavation can be presented according to Equations (10) and (11):

$$E_{ME} = E_R + E_E + E_{OE} \tag{10}$$

$$E_{ME} = \sum_{mi} \left(\eta_{mi} \cdot \underline{V_{FO_{mi}}} \cdot d_{FOmi} \cdot CV_{FOmi} \right) \tag{11}$$

In the above equations, the following definitions are used:

E_{ME} is a total sum of energy consumption from mechanical excavation, MJ;
E_R is the energy consumption from ripping operations by dozer, MJ;
E_E is the energy consumption from excavation operations by a backhoe with a skidder, MJ;

E_{OE} is the energy consumption from oversize excavation operations by a backhoe with a skidder, MJ;

η_{mi} is the total efficiency of the powertrain of all machines in the technological system, %;

$\underline{V_{FOmi}}$ is the average volume of combusted diesel oil in $dm^3 \cdot h^{-1}$;

d_{FOmi} is a fuel oil density, $dm^3 \cdot h^{-1}$;

CV_{FOmi} is a calorific value of combusted fuel oil in $MJ \cdot kg^{-1}$.

According to Table 3, the average fuel consumption for the Caterpillar D9R crawler dozer and the Caterpillar 330DL excavator is 26.76 $dm^3\ h^{-1}{}_1$ and 22.76 $dm^3 \cdot h^{-1}$, respectively. Considering the work hours needed to excavate 12,779 tons of gypsum (assuming only excavation with no additional maneuvering) and the average fuel consumption of each machine, the total diesel fuel required amounts to 6781.38 kg (3285.27 kg for Caterpillar 330DL and 3496.10 kg for Caterpillar D9R).

Given the average density of diesel fuel at 0.8325 $kg \cdot dm^3$ and a diesel calorific value of 42.6 $MJ \cdot kg$, the total energy consumption for Caterpillar 330DL, factoring in its transmission system efficiency of 35%, was 40,778.71 MJ. For Caterpillar D9R, the energy consumption, based on a diesel consumption of 3496.10 kg and a transmission efficiency of 37%, amounted to 45,875.43 MJ. Consequently, the total energy consumption for mechanical excavation operations (ripping and quarrying) was 86,654.15 MJ, as shown in Table 10 and Figure 9.

Table 10. Average energy consumption of each mechanical excavation operation equivalent to the gypsum output from one blasting series.

Operation	Energy Consumption, MJ
E_R—Ripping operation by dozer	45,875.4
E_E—Excavation operation by excavator with excentric skidder	40,778.7
E_{OE}—Oversize excavation operation by excavator with excentric skidder	86,654.2

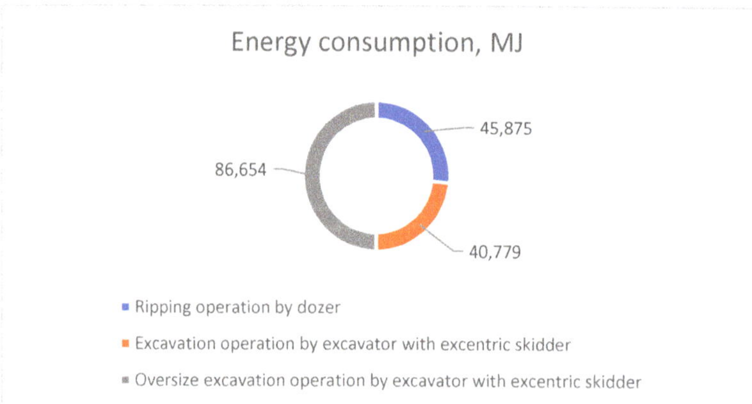

Figure 9. Energy consumption during mechanical extraction.

3.2.2. Evaluation of the Fumes Generated by Mechanical Excavation

The emissions of NOx and CO resulting from diesel fuel combustion during mechanical excavation were evaluated using the parameters defined in [40] and according to Equation (11). The analysis considered both gross power and operational hours for the Caterpillar D9R dozer and the Caterpillar 330DL hydraulic excavator. For Caterpillar D9R, the base fume emissions were estimated at 543.6 $g\ h^{-1}$ for NOx and 450 $g\ h^{-1}$ for CO. With an average operational time of 130.6 h, this resulted in emissions of approximately 71.00 kg of NOx and 58.78 kg of CO (Table 11). Additionally, the combustion of 1 dm^3 of diesel fuel emits approximately 2.67 kg of CO_2, leading to an estimated CO_2 emission of 9334.60 kg for the dozer during mechanical excavation. For the Caterpillar 330DL excavator, base emissions were estimated at 648 g h^{-1} for NOx and 300 g h^{-1} for CO. With an operational time of approximately 144.4 h, this resulted in emissions of approximately 93.56 kg of CO and 43.31 kg of NOx. The corresponding CO_2 emissions were estimated at 8771.68 kg. Overall, the mechanical extraction of 12,779 tons of ore resulted in emissions of approximately 18,106.28 kg of CO_2, 152.34 kg of CO, and 114.32 kg of NO_x, as shown in Table 11 and Figure 10.

Table 11. Average fume emissions from blasting mechanical excavation in this case study.

Type of Operation	Amount, kg		
	CO_2	CO	NO_x
Ripping process	9334.60	58.78	71.00
Excavation operations	8771.68	93.56	43.31
Overall	18,106.28	152.33	114.32

(**a**)

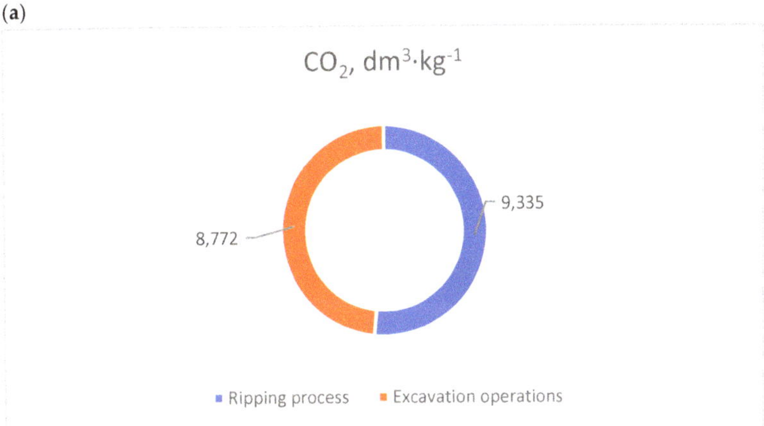

(**b**)

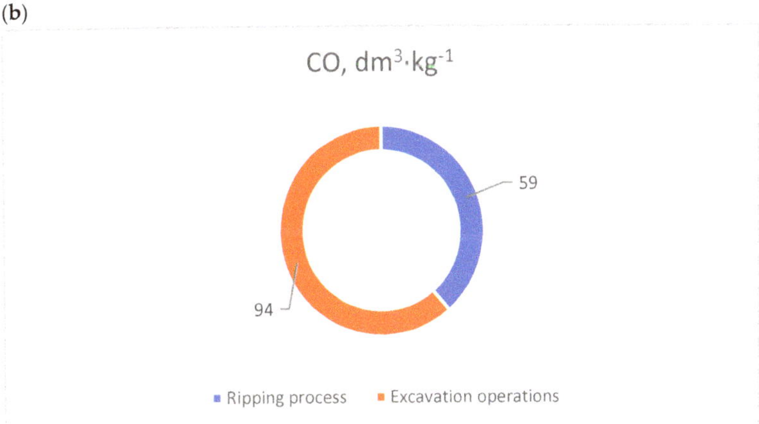

Figure 10. *Cont.*

(**c**)

(**d**)

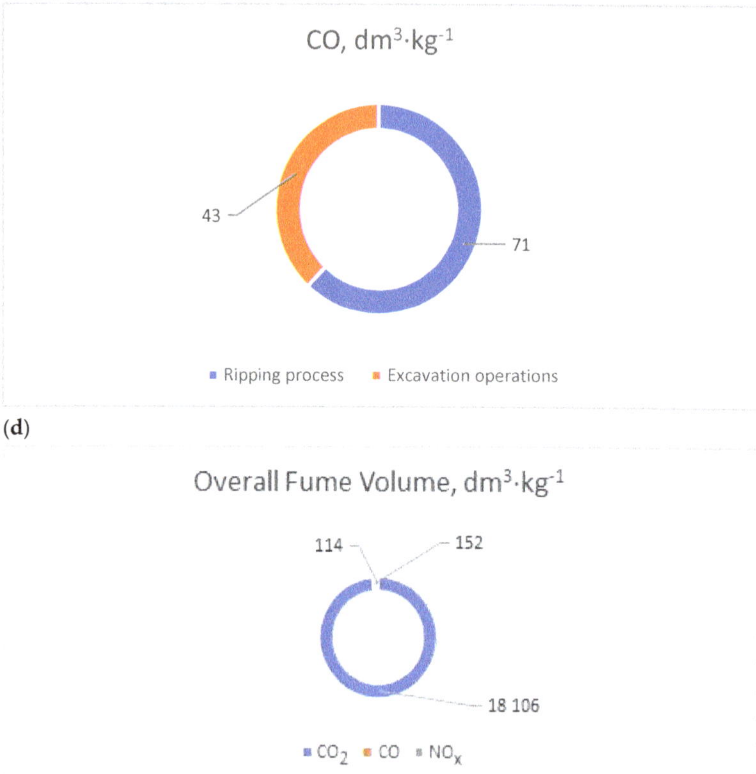

Figure 10. Total fumes emitted from all mechanical extraction operations: (**a**) CO, (**b**) CO_2, (**c**) NO_x, and (**d**) overall.

3.3. Comparison of Energy Consumption and Gas Emissions

The total energy consumption for drilling and blasting ranged from 1298.71 to 1452.58 MJ, as shown in Table 6, and, for mechanical mining, it equaled 173,308.30 MJ, as shown in Table 9. Despite the application of high-energy agents, the amount of energy required to extract the deposit was more than 100 times less than for mechanical excavation because of the accumulation of energy from the detonation in the mining process. For drilling and blasting, the most energy-consuming operation was borehole drilling (45.8% of the total energy), and the explosion itself consumed only 21.0% of the total energy of the extraction process. For the mechanical mining case, the energy required to extract the rock from the deposit was roughly equal to the energy consumption of re-crushing excessively large boulders produced during the extraction process. This is because properly adjusted blasting parameters allow one to obtain homogenous fragmentation of the extracted rock.

To compare the discussed mining techniques in terms of energy consumption, it was necessary to relate the impact of effective mining energy to the unit volume of the extracted deposit. Figure 11 shows the comparison of the energy intensity of the mining process and the effective productivity per hour for the analyzed mining techniques.

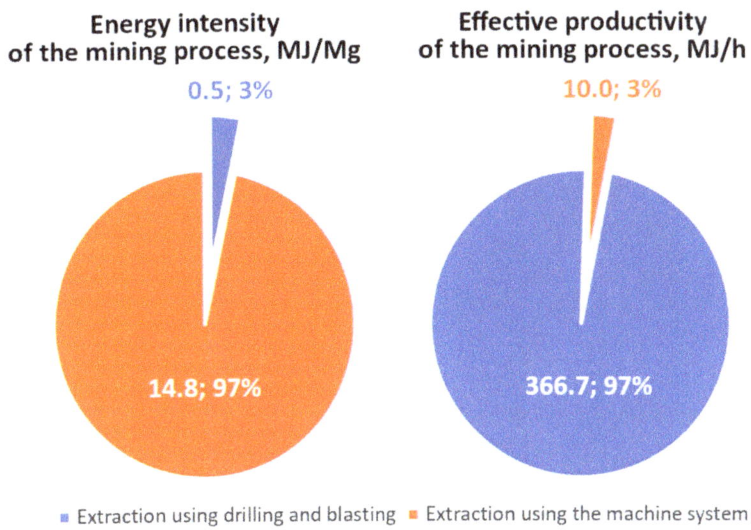

Figure 11. Relation of energy intensities and effective productivities of the mining process using blasting works and mechanical extracting.

As can be seen from the analysis carried out for the process of mining with explosives and mechanical mining, these modes present an inversely proportional relationship regarding the energy required to mine part of the same deposit, Figure 11.

The total emission of fumes for drilling and blasting is 305.43 kg, as shown in Table 9, and, for mechanical mining, this equals 18,372.93 kg, as shown in Table 10. Although explosives produce an enormous amount of gaseous products per unit of mass, it has been found that the total amount of fumes produced in this process is more than six times less than for mechanical mining.

In both cases, carbon dioxide emissions are the highest. Considering blasting works, more nitrogen oxides are produced than carbon oxides, while the reverse applies to mechanical mining. It is evident that mechanical extraction results in significantly higher emissions, primarily due to the greater volume of diesel combusted. However, during the prolonged mechanical extraction process, these emissions are gradually dispersed into the environment.

The blasting technique, which involves drilling and explosive loading, results in the instantaneous release of high concentrations of fumes from explosive decomposition. This can pose significant occupational hazards to employees. However, despite being more energy-intensive (86,654.15 MJ) than blasting (1298.12 to 1462.22 MJ), mechanical extraction does not generate adverse environmental effects such as air blasts, blast-induced vibrations, or fly rocks. The absence of these effects is a major advantage of mechanical extraction. Nonetheless, it should be noted that the efficiency of mechanical extraction may decrease over time due to the potential wear and tear on technical equipment.

3.4. Prediction Model Results

The prediction of the total extraction volume was made using a 5-year dataset of extraction records from the mine, as shown in Table 1.

In Table 12, a gradual decline in extraction volume can be observed, decreasing from approximately 800,000 tons to around 530,000 tons, with a slight recovery to 600,000 tons in the final year. Based on these trends, it is anticipated that extraction volumes will continue to decrease, as shown in Figure 12, due to the depletion of the deposit at the current operational levels of the open-pit mine. The opening of a new operational level or the development of a new excavation area could alter this predicted trend.

Table 12. Overall annual extraction, energy consumption with the assumption of 80% energy dissipation from explosive detonation, and fume emission from blasting operation and mechanical extraction.

Year	Gypsum Mass, Tons	Energy Consumption, MJ	Amount, kg		
			CO_2	CO	NO_x
1	804,473	949,506	192,902.6	3398.6	3934.1
2	827,956	1,031,034	209,789.4	3588.6	4138.6
3	529,479	735,154	150,017.3	2422.8	2772.9
4	529,320	961,798	176,508.9	2636.1	2983.2
5	619,790	552,251	111,117.9	2315.9	2732.5

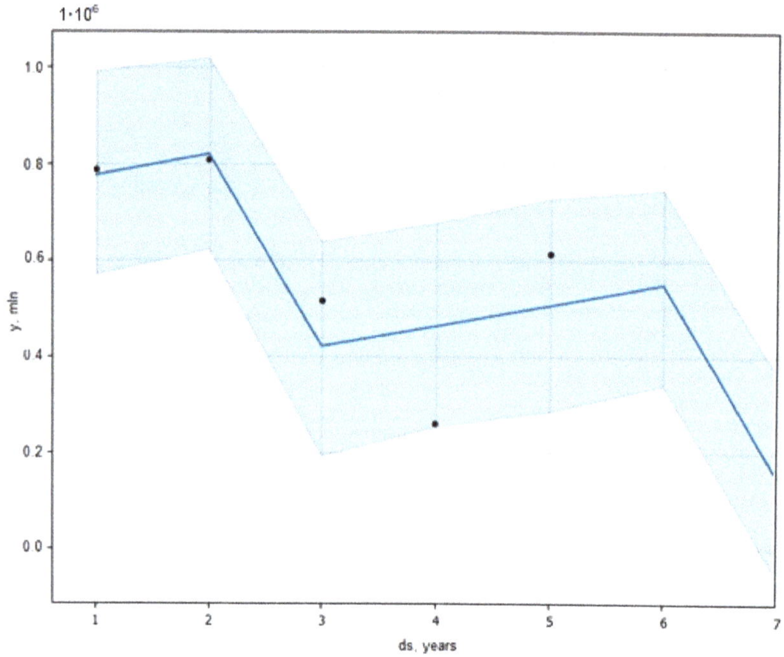

Figure 12. Prediction model for overall gypsum extraction.

The reduction in gypsum extraction volume will be accompanied by a corresponding decrease in energy consumption, as depicted in Figure 13.

However, it should be noted that, when extraction volumes decrease, many fixed energy costs remain constant, leading to a higher energy intensity since the same amount of energy is used to extract a smaller volume of ore. This scenario applies only if the extracted gypsum volume continues to diminish. Should the open-pit mine initiate a new operational level or open a new excavation area, energy consumption related to haulage and preparatory works (which are not accounted for in the current analysis) would increase.

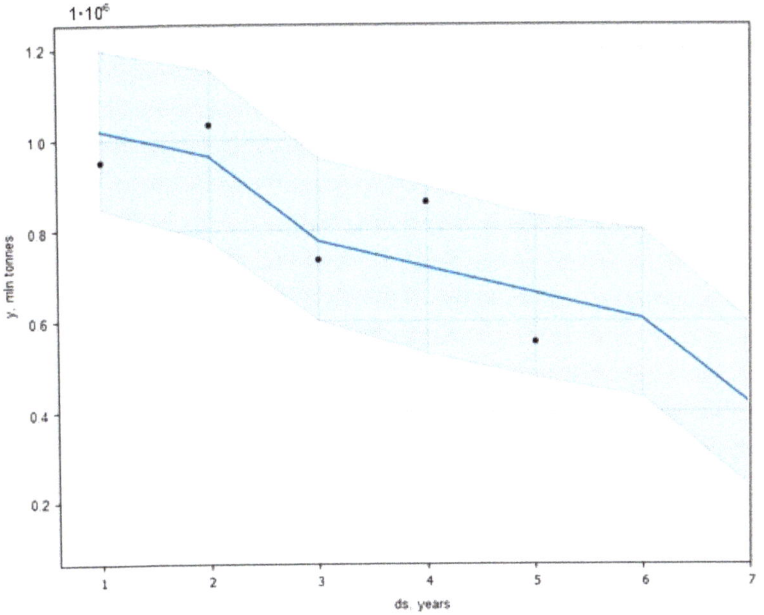

Figure 13. Prediction model of overall energy consumption.

As the open-pit mine deepens, the distance between the pit and the processing facility increases, resulting in longer haulage times. This increase in haulage distance necessitates greater fuel consumption by trucks, directly affecting the energy efficiency of the operation. Additionally, mine expansion requires extensive preparatory work, including the removal of overburden to access the deposit. The increased volume of waste material to be removed and transported adds to the energy demand, reducing the efficiency of ore extraction and raising the energy consumption per ton of ore mined. These factors must be considered when evaluating the implementation of new extraction options. A decrease in energy consumption would also lead to a reduction in emissions, due to lower demands for fuel and explosives in mechanical extraction and blasting operations. The predictive model developed for carbon dioxide, carbon monoxide, and nitrogen oxide emissions corroborates this, as shown in Figure 14. However, if the total volume of extracted gypsum were to increase, the emission volume would rise proportionally as well.

(a)

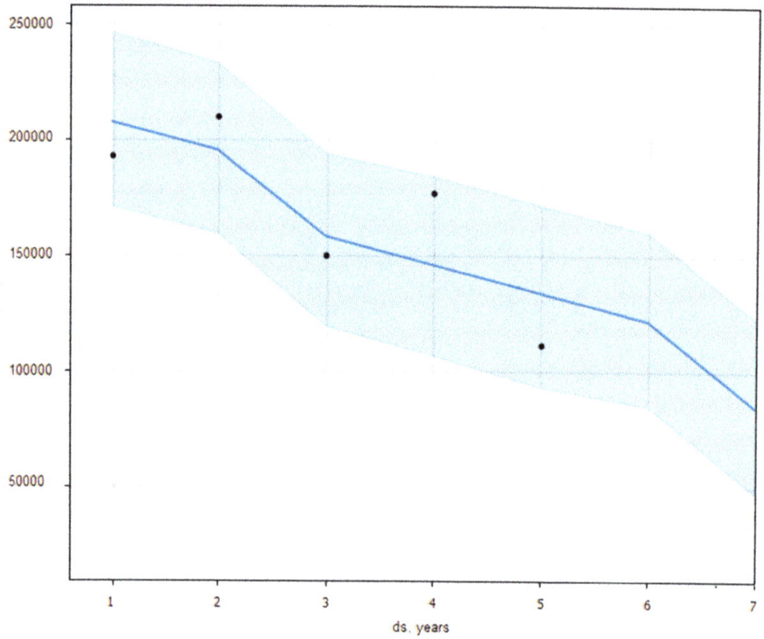

(b)

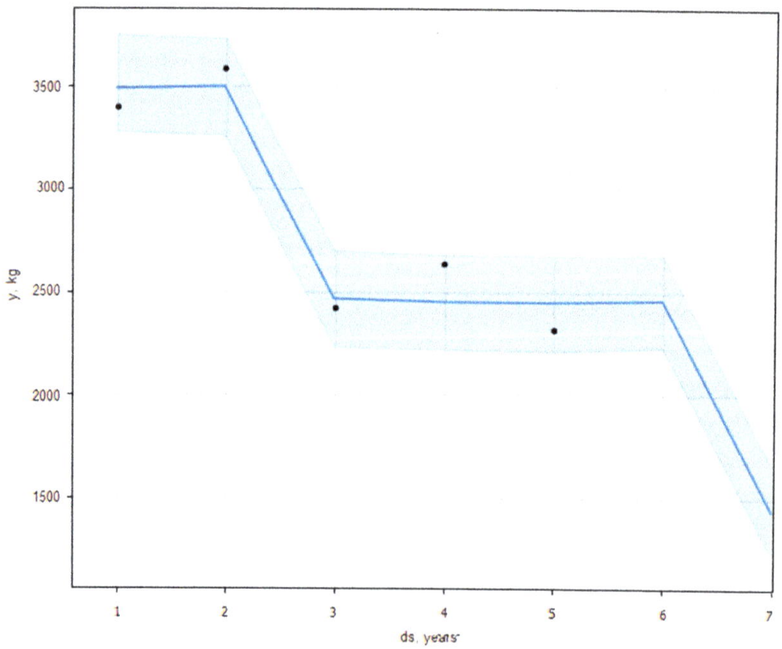

Figure 14. *Cont.*

(c)

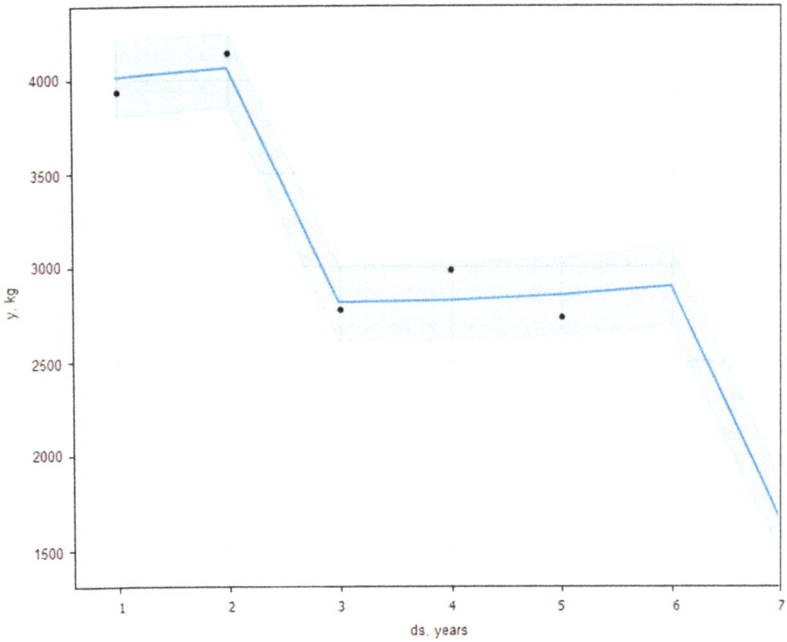

Figure 14. Prediction models for the emission of (**a**) CO_2, (**b**) CO, and (**c**) NO_x.

3.5. Theoretical Discussion on Environmental Impact

As mentioned in the Introduction Section, quarrying can have a negative impact on the environment and local communities, regardless of the method of extraction [17,36,49–52]. Current studies show that the most common impacts of blasting operations in surface mines include blast-induced vibration [53–55], fly rock [56–58], air overpressure [59–61], air dust [62–64], noise [65–67], and gaseous emissions [68,69]. In the process of drilling blastholes with drilling rigs, noise, gas emissions from diesel engines, and air dust emissions can be identified [70,71]. Machine-based quarrying mainly concerns itself with the effects of noise, dust emission, and gas pollution from diesel engines [72,73]. The impact of quarry operations can be considered in terms of the nature and extent of the impact. Quarrying may affect the natural environment (i.e., greenhouse gas emission or noise), the anthropogenic environment (vibrations' effect on the technical condition of nearby buildings), and human health (i.e., silica exposure from rock dust or noise exposure). Depending on the nature of the hazard, the impacted area can vary. The environmental impacts of blasting operations, depending on technological factors and mining and geological conditions, may affect the distant surroundings of the mine (blast-induced vibration, air overpressure), while others are rather local, limited to the direct vicinity. In mechanical quarrying, the most far-reaching impact is noise, while other hazards are mostly confined to the pit area. Due to the impact zone, limited mitigation options, and the urbanized surroundings of quarries, blast-induced vibrations and air overpressure are major sources of risk in quarrying [74–76]. Drilling and blasting may not be feasible in sites with short distances between the mine face and buildings or where geological conditions are unfavorable. In such cases, one of the most common approaches to ensuring production is the use of mechanical-based rock excavation.

4. Conclusions

This manuscript presents the comparison of energy consumption for two mining methods (gypsum extraction with blasting techniques and mechanical extraction).

The energy consumption analysis indicated that mechanical extraction with a combination of a crawler excavator and a crawler bulldozer required much more energy (86,654.15 MJ) compared to blasting techniques. In the case of blasting techniques, energy consumption ranged from 1298.12 to 1462.22 MJ. This range depended on the effective utilization of the detonation energy for quarrying. Moreover, approximately 970 MJ of energy was obtained from the detonation process. The residual energy was derived from diesel combustion in explosive loading and drilling operations. The energy from diesel combustion in both mining systems dissipated over a long period of time, resulting in the dilution of fumes. Based on the obtained data, it was established that the extraction of 12,779 tons required, depending on the extraction system, approximately 7.46 h (blasting techniques) or 130.6 h (mechanical excavation). This reveals that, in terms of operation time, the blasting technique is much more efficient. The analysis showed that, while blasting is more energy-efficient in terms of energy used per unit of extracted material, the mechanical excavation efficiency may decrease over time due to equipment wear and tear. This decline in efficiency can impact the overall productivity of the mining process.

Although explosives are classified as high-energy materials, the extended duration of mineral extraction for an equivalent volume of gypsum resulted in mechanical extraction being a significantly more energy-intensive process. This was primarily due to the substantial amount of fuel required for combustion during mechanical operations. Moreover, the energy required for mechanical extraction was released with time.

The fume analysis indicated that, in terms of blasting operations, the CO and NOx are mostly generated by the explosive's detonation. The CO_2 is mostly derived from the combustion of diesel. However, the composition of the fumes may be influenced by the type of explosives that are used. In terms of mechanical extraction, the overall emission of fumes was 18,106.28 kg of CO_2, 152.34 kg of CO, and 114.32 kg of NO.

The prediction model indicated that gypsum extraction will decrease significantly within a relatively short period unless proactive measures, such as opening a new operational level or expanding the mine, are undertaken. Based on the data obtained and analyzed using the Prophet algorithm, it was established that the reduction in extraction volume will be accompanied by a corresponding decrease in both energy consumption and emission of fumes. However, it is important to note that, as extraction volumes decline, many fixed energy costs, such as those associated with machinery operation and mine maintenance, will remain constant. This will result in an increase in energy intensity, as the same amount of energy will be required to maintain operations, despite the smaller output of extracted material. Consequently, energy efficiency will deteriorate as the energy consumption per unit of extracted deposit will increase.

The FBProphet algorithm offers a robust and accessible approach for the predictive analysis of diverse trends. This algorithm has never been used before in the mining industry. The accuracy of the model is largely dependent on the dataset size, with larger datasets typically enhancing model precision. Additionally, the algorithm is particularly well-suited for time series modeling involving multiple seasonal patterns, making it a valuable tool for predicting general extraction volumes and other industry-relevant metrics. As the analysis showed, apart from the need for the mechanical excavation of hard rock, the impact of the detonation of explosives in the form of a blast wave, and the scattering range of rock fragments, the only factor that can be considered is cost. The authors plan to add an economic aspect to the case study discussed here in a future analysis.

Author Contributions: Conceptualization, A.B., M.D. and P.B.; methodology, A.B. and P.B.; validation, M.D.; formal analysis, A.B., P.B. and M.D.; investigation, A.B., M.D. and P.B.; resources, P.B.; data curation, A.B., M.D. and P.B.; writing—original draft preparation, A.B., P.B. and M.D.; writing—review and editing, A.B., M.D. and P.B.; visualization, M.D.; and supervision, A.B. All authors have read and agreed to the published version of the manuscript.

Funding: The authors wish to thank the Faculty of Civil Engineering and Resource Management at the AGH University of Krakow for the financial support for this research (no. 16.16.100.215).

Institutional Review Board Statement: Not applicable.

Informed Consent Statement: Not applicable.

Data Availability Statement: The original contributions presented in this study are included in the above article; further inquiries can be directed to the corresponding author.

Acknowledgments: The authors extend their gratitude to the management of Saint-Gobain Construction Products Polska Sp. z o.o. Special acknowledgment is due to Mateusz Gajda, Mine Manager, for his invaluable assistance with the mechanical extraction and blasting works at the Saint-Gobain Construction Products facility in Szarbków, Poland.

Conflicts of Interest: The authors declare no conflicts of interest.

References

1. Machniak, Ł.; Kozioł, W.; Borcz, A. Wytyczne wyboru efektywnych układów wydobywczych do produkcji kruszyw łamanych. *Górnictwo Odkryw.* **2013**, *54*, 95–101.
2. Zou, D. Mechanisms of Rock Breakage by Blasting. In *Theory and Technology of Rock Excavation for Civil Engineering*; Springer: Singapore, 2016.
3. Hamdi, E.; Romdhane, N.B.; Mouza, J.; Le Cleach, J.M. Fragmentation Energy in Rock Blasting. *Geotech. Geol. Eng.* **2007**, *26*, 133–148. [CrossRef]
4. Lusk, B.; Silva, J.J. Energy Distribution in the Blast Fragmentation Process. In *Energy Efficiency in the Minerals Industry. Green Energy and Technology*; Awuah-Offei, K., Ed.; Springer: Berlin/Heidelberg, Germany, 2017.
5. Pyra, J. Impact assessment of mining blasting operations for buildings. *WUG Bezpieczeństwo Pracy i Ochrona Środowiska w Górnictwie* **2008**, *3*, 41–47.
6. Sanchidrián, J.A.; Segarra, P.; López, L.M. Energy components in rock blasting. *Int. J. Rock Mech. Min. Sci.* **2007**, *44*, 130–147. [CrossRef]
7. Armaghani, J.D.; Hajihassani, M.; Monjezi, M.; Tonnizam, M.; Marto, A.; Moghaddam, M.R. Application of two intelligent systems in predicting environmental impacts of quarry blasting. *Arab. J. Geosci.* **2015**, *8*, 9647–9665. [CrossRef]
8. Fişne, A.; Kuzu, C.; Hüdaverdi, T. Prediction of environmental impacts of quarry blasting operation using fuzzy logic. *Environ. Monit. Assess.* **2011**, *174*, 461–470. [CrossRef]
9. Hidayat, S. Environmental impacts of open pit mining blasting: Particular discussions on some specific issues. *J. Min. Environ.* **2021**, *1*, 1–11.
10. Jung, J.; Grabar, K. Methods for reducing the environmental impact of rock mass excavation. *EnvEng-IO* **2020**, *7*, 29–38.
11. Kuzu, C.; Ergin, H. An assessment of environmental impacts of quarry-blasting operation: A case study in Istanbul, Turkey. *Environ. Geol* **2005**, *48*, 211–217. [CrossRef]
12. Tomberg, T.; Toomik, A. Environmental impact of mine blasting. *Environ. Technol. Resour. Proc. Int. Sci. Pr. Conf.* **1999**, *1*, 213–219. [CrossRef]
13. Biessikirski, A.; Dworzak, M.; Twardosz, M. Composition of Fumes and Its Influence on the General Toxicity and Applicability of Mining Explosives. *Mining* **2023**, *3*, 605–617. [CrossRef]
14. Goswami, T.; Brent, G. Blasting approaches to increase mine productivity and reduce greenhouse gas emissions in surface coal mining. In Proceedings of the 11th International Symposium on Rock Fragmentation by Blasting, Sydney, Australia, 24–26 August 2015; pp. 635–644.
15. Kasztelewicz, Z.; Zajączkowski, M.; Sikora, M. Przegląd mechanicznych sposobów urabiania skał zwięzłych. *Pr. Nauk. Inst. Górnictwa Politech. Wrocławskiej* **2013**, *43*, 85–98.
16. Kittipongvises, S. Assessment of Environmental Impacts of Limestone Quarrying Operations in Thailand. *Environ. Clim. Technol.* **2017**, *20*, 67–83. [CrossRef]
17. Langer, W. *Potential Environmental Impacts of Quarrying Stone in Karst—A Literature Review*; U.S. Geological Survey Report; U.S. Department of the Interior, USGS Publishing Warehouse: Reston, VA, USA, 2001.
18. Toraño, J.; Rodriguez, R.; Diego, I.; Menéndez, M. Environmental impact of rock excavation in urban areas: Comparison between blasting and hydraulic breaker hammer. *Civ. Eng. Environ. Syst.* **2006**, *23*, 117–126. [CrossRef]

19. Amar, P.; Ramachandra, M.V.M.S.; Bahadur, S.K. Rock excavation using surface miners: An overview of some design and operational aspects. *Int. J. Min. Sci. Technol.* **2013**, *23*, 33–40.
20. Ismael, M.; Abdelghafar, K.; Sholqamy, M.; Elkarmoty, M. Performance prediction of hydraulic breakers in excavation of a rock mass. *Rud. Geološko-Naft. Zb. (Min. Geol. Pet. Eng. Bull.)* **2021**, *36*, 107–119. [CrossRef]
21. Pebrianto, R.; Asof, M.; Susilo, B.K.; Gofar, N. Evaluation of Factors Affecting Ripping Productivity in Open Pit Mining Excavation. *Electron. J. Geotech. Eng.* **2014**, *19*, 10447–10456.
22. Cheenkachorn, K.; Poompipatpong, C.; Ho, C.G. Performance and emissions of a heavy-duty diesel engine fuelled with diesel and LNG (liquid natural gas). *Energy* **2013**, *53*, 52–57. [CrossRef]
23. Huang, W.D.; Zhang, Y.H.P. Energy efficiency analysis: Biomass-to-wheel efficiency related with biofuels production, fuel distribution, and powertrain systems. *PLoS ONE* **2011**, *6*, e22113. [CrossRef]
24. Kanfar, M.; Korman, T.; Kujundžić, T. Fuel consumption and engine load factors of equipment in quarrying of crushed stone. *Teh. Vjesn.* **2016**, *23*, 163–169.
25. Okamoto, T.; Uchida, N. New Concept for Overcoming the Trade-Off between Thermal Efficiency, Each Loss and Exhaust Emissions in a Heavy Duty Diesel Engine. *SAE Int. J. Engines* **2016**, *9*, 859–867. [CrossRef]
26. Ismail, M.A.M.; Kumar, N.S.; Abidin, M.H.Z.; Madun, A. Rippability Assessment of Weathered Sedimentary Rock Mass using Seismic Refraction Methods. *J. Phys. Conf. Ser.* **2018**, *995*, 012105. [CrossRef]
27. Bęben, A. *Teoretyczne Podstawy Mechanicznego Zwiercania Skał w Górnictwie Odkrywkowym*; Wydawnictwa AGH: Kraków, Poland, 2012.
28. Zhu, J.X.; Zhao, C.Y.; Zou, X.F. Identification of soil parameters on vibratory excavation of hydraulic excavator. *Zhongnan Daxue Xuebao (Ziran Kexue Ban)/J. Cent. South Univ. (Sci. Technol.)* **2006**, 37.
29. Zhu, J.X.; Zhao, C.Y.; Guo, X. Research on mechanism of vibratory excavation of hydraulic excavators. *Yantu Lixue/Rock Soil Mechanics.* **2007**, *28*, 537–541.
30. Zhu, J.; Hu, H.; Mei, Y. Power consumption study on bucket vibratory excavation of hydraulic excavator. *Tongji Daxue Xuebao/J. Tongji Univ.* **2008**, *36*, 1408–1412.
31. Boudiaf, M.; Chaib, R.; Talhi, K. Study of rippability in limestone quarries-case of the hadjar-soud quarry, Algeria. *Phys. Chem. News* **2010**, *53*, 39–44.
32. Stewart, P.; Ramezanzadeh, A.; Knights, P. Benchmark Drill and Blast and Mechanical Excavation Advance Rates for Underground Hard-Rock Mine Development. In Proceedings of the Australian Mining Technology Conference, Hunter Valley, NSW, Australia, 26–27 September 2006; pp. 41–63.
33. Wennmohs, K.H. Mechanical Rock Excavation versus Drilling and Blasting. *GeoRecources J. Resour. Min. Tunn. Geotech. Equip.* **2021**, *2*, 32–36.
34. Feijóo, P.C. Comparative Analysis Between Drill and Blast and Mechanical Excavation in Underground Construction. Master's Thesis, National Technical University of Athens, School of Mining and Metallurgical Engineering, Athens, Greece, 2019.
35. Skawina, B.; Greberg, J.; Salama, A.; Schunnesson, H. Mechanical Excavation and Drilling and Blasting—A Comparison Using Discrete Event Simulation. *Mine Plan. Equip. Sel.* **2014**, 367–377.
36. Bascompta, M.; Sanmiquel, L.; Gangolells, M.; Sidki, N. LCA analysis and comparison in quarrying: Drill and blast vs mechanical extraction. *J. Clean. Prod.* **2022**, *369*, 133042. [CrossRef]
37. *EN 13631-16:2004*; Explosives for Civil Uses. High Explosives. Part 16: Detection and Measurement of Toxic Gases. European Committee for Standardization: Brussels, Belgium, 2004.
38. European Union. Council directive 93/15/EEC of 5 April 1993 on the harmonisation of the provisions relating to the placing on the market and supervision of explosives for civil uses. *Off. J. Eur. Union* **1993**, *12*, 20–36.
39. Sadowski, A.; Żółtowski, B. Badania sprawności złożonych układów napędowych. *Inżynieria I Apar. Chem.* **2012**, *5*, 249–250.
40. Winther, M.; Dore, C. EMEP/EEA Air Pollutant Emission Inventory Guidebook. 2023. Available online: https://www.eea.europa.eu/publications/emep-eea-guidebook-2023 (accessed on 1 August 2024).
41. Harvey, A.C.; Shephard, N. Structural time series models. In *Handbook of Statistics*; Maddala, G., Rao, C., Vinod, H., Eds.; Elsevier: Amsterdam, The Netherlands, 1993; Volume 11, pp. 261–302.
42. Harvey, A.C.; Peters, S. Estimation procedures for structural time series models. *J. Forecast.* **1990**, *9*, 89–108. [CrossRef]
43. Taylor, S.J.; Letham, B. Forecasting at Scale. *Am. Stat.* **2018**, *72*, 37–45. [CrossRef]
44. OZN Research. DCA 5 Detonation Calorimeter Product Datasheet. Available online: https://mueller-instruments.de/fileadmin/Downloads/instruments-medien/Detonationskalorimeter_-_DCA_5.pdf (accessed on 1 August 2024).
45. Rai, R.; Singh, T.N. A new predictor for ground vibration prediction and its comparison with other predictors. *Indian J. Eng. Mater. Sci.* **2004**, *11*, 178–184.
46. Samujłło, J. *Roboty Strzelnicze w Górnictwie Odkrywkowym*; Wydawnictwo Śląsk: Katowice, Poland, 1968.
47. Orlen Poludnie. Ekoterm Product Datasheet. Available online: https://www.orlenpoludnie.pl/PL/NaszaOferta/StrefaBIOpaliw/OlejeOpalowe/Strony/Ekoterm.aspx (accessed on 9 August 2024).
48. United States Environmental Protection Agency. Euxahust Emissions Calculations. Available online: https://www.epa.gov/sites/default/files/2015-09/documents/nghwemit.pdf (accessed on 9 August 2024).

49. Lee, C.; Asbjörnsson, G.; Hulthén, E.; Evertsson, M. The environmental impact of extraction: A holistic review of the quarry lifecycle. *Clean. Environ. Syst.* **2024**, *13*, 100201. [CrossRef]
50. Omosanya, K.O.; Ajivade, O.M. Environnmental impact of quarrying on Otere Village, Odeda, Southwestern Nigeria. *Ozean J. Appl. Sci.* **2011**, *4*, 75–82.
51. Capitano, C.; Peri, G.; Rizzo, G.; Ferrante, P. Towards a holistic environmental impact assessment of merble quarrying and processing: Proposal of a novel easy-to-use IPAT-based method. *Environ. Monit. Assess* **2017**, *189*, 108. [CrossRef]
52. Domingues, J.M.; Miranda, V.F.; Rezende, D.C.; Lares, Y.S.; Ferreira, S.R.; de Oliveira, I.R. Statistical modeling of qyarrying activities and their impact on residents' satisfaction. *J. Environ. Sci. Sustain. Dev.* **2020**, *3*, 416–429. [CrossRef]
53. Feher, J.; Cambal, J.; Pandula, B.; Kondela, B.; Sofranko, M.; Mudarri, T.; Buchla, I.; Feher, J.; Cambal, J.; Pandula, B.; et al. Research of the technical seismicity due to blasting works in quarries and their impact on the environment and population. *Appl. Sci.* **2021**, *11*, 2118. [CrossRef]
54. Sevelka, T. Blasting Quarry Operations: Adverse and Cumulative Effects, Lawsuits and Complaints, and Suggested Remedies. *J. Envtl. L. Pol'y* **2023**, *3*, 1–79. [CrossRef]
55. Djaksimuratov, K.; Bayramova, M. Environmental impact of rock blasting processes in mining enterprises. *Mod. Sci. Res.* **2024**, *6*, 842–847.
56. Nikafshan Rad, H.; Bakhshayeshi, I.; Wan Jusoh, W.A.; Tahir, M.M.; Foong, L.K. Prediction of flyrock in mine blasting: A new computational intelligence approach. *Nat. Resour. Res.* **2020**, *29*, 609–623. [CrossRef]
57. Murlidhar, B.R.; Nguyen, H.; Rostami, J.; Bui, X.; Armaghani, D.J.; Ragam, P.; Mohamad, E.T. Prediction of flyrock distance induced by mine blasting using a novel Harris Hawks optimization-based multi-layer perceptron neural network. *J. Rock Mech. Geotech. Eng.* **2021**, *13*, 1413–1427. [CrossRef]
58. Bhatawdekar, R.M.; Kumar, R.; Sabri Sabri, M.M.; Roy, B.; Mohamad, E.T.; Kumar, D.; Kwon, S. Estimating flyrock distance induced due to mine blasting by extreme learning machine coupled with an equilibrium optimizer. *Sustainability* **2023**, *15*, 3265. [CrossRef]
59. Nguyen, H.; Bui, X.N.; Tran, Q.H. Estimating air over-pressure resulting from blasting in quarries based on a novel ensemble model (GLMNETs–MLPNN). *Nat. Resour. Res.* **2021**, *30*, 2629–2646. [CrossRef]
60. Fang, Q.; Nguyen, H.; Bui, X.N.; Tran, Q.H. Estimation of blast-induced air overpressure in quarry mines using cubist-based genetic algorithm. *Nat. Resour. Res.* **2020**, *29*, 593–607. [CrossRef]
61. Harandizadeh, H.; Armaghani, D.J. Prediction of air-overpressure induced by blasting using an ANFIS-PNN model optimized by GA. *Appl. Soft Comput.* **2021**, *99*, 106904. [CrossRef]
62. Zvyagintseva, A.V.; Sazonova, S.A.; Kulneva, V.V. Analysis of sources of dust and poisonal gases in the atmosphere formed as a result of explosions at quarries of the mining and integrated works. *Proc. IOP Conf. Ser. Mater. Sci. Eng.* **2020**, *962*, 042045. [CrossRef]
63. Normatova, M.; Pardayeva, S. Reducing dust pollution in the production of mass explosions in quarries. *Proc. EDP Sci. E3S Web Conf.* **2024**, *525*, 01003. [CrossRef]
64. Bakhtavar, E.; Hosseini, S.; Hewage, K.; Sadiq, R. Air pollution risk assessment using a hybrid fuzzy intelligent probability-based approach: Mine blasting dust impacts. *Nat. Resour. Res.* **2021**, *30*, 2607–2627. [CrossRef]
65. Aksoy, M.; Hakan, A.; Konuk, A. Development of a Preliminary Blasting Design and Assessment of Environmental Impacts for a Quarry. *Çukurova Üniversitesi Mühendislik-Mimar. Fakültesi Derg.* **2019**, *34*, 241–248. [CrossRef]
66. Kumar, A.; Prasad, S.; Reddy, R.S. Environmental implications on blasting operations in Indian quarry mines. *Int. J. Min. Geo-Eng.* **2024**, *58*, 289–294.
67. Hassan, I.A. Impact of quarry activities on the serenity of the neighbouring localities in Ogun State, Nigeria. *Afr. J. Agric. Technol. Environ.* **2022**, *11*, 30–37.
68. Al-Bakri, A.; Hefni, M. A review of some nonexplosive alternative methods to conventional rock blasting. *Open Geosci.* **2021**, *13*, 431–442. [CrossRef]
69. Eruke, O.S.; Igwenagu-Ifeanyi, V.; Belonwu, D.C. Assessment of drinking water and air quality around selected quarries in southeastern Nigeria. *World J. Innov. Res.* **2021**, *11*, 70–76.
70. Regotunov, A.; Sukhov, R.; Bersenyov, G. About transition processes in blasthole drilling at quarries. *Proc. EDP Sci. E3S Web Conf.* **2020**, *177*, 01008. [CrossRef]
71. Ganapathi, H.; Phukan, M. Environmental hazards of limestone mining and adaptive practices for environment management plan. In *Environmental Processes and Management. Water Science and Technology Library*; Singh, R., Shukla, P., Singh, P., Eds.; Springer: Cham, Switzerland, 2020; Volume 91, pp. 121–134. [CrossRef]
72. Gupta, P.; Roy, S.; Babu, A.R. Study on noise levels generated due to jack hammer drills in granite quarries. *Front. Sci.* **2012**, *2*, 47–52. [CrossRef]
73. Singhal, A.; Goel, S. Impact of Sandstone Quarrying on the Health of Quarry Workers and local residents: A case study of Keru, Jodhpur, India. In *Treatment and Disposal of Solid and Hazardous Wastes*; Springer: Cham, Switzerland, 2022; pp. 97–118.
74. Toraño, J.; Ramírez-Oyanguren, P.; Rodríguez, R.; Diego, I. Analysis of the environmental effects of ground vibrations produced by blasting in quarries. *Int. J. Min. Reclam. Environ.* **2006**, *20*, 249–266. [CrossRef]

75. Spathis, A.T. A brief review of the measurement, modelling and management of vibrations produced by blasting. In Proceedings of the 9th International Symposium on Rock Fragmentation by Blasting, Granada, Spain, 13–17 August 2009; pp. 1–11.
76. Zeng, J.; Mohammed, A.S.; Mirzaei, F.; Moosavi, S.M.H.; Armaghani, D.J.; Samui, P. A parametric study of ground vibration induced by quarry blasting: An application of group method of data handling. *Environ. Earth Sci.* **2022**, *81*, 127. [CrossRef]

energies

MDPI

Article

Potentials of Green Hydrogen Production in P2G Systems Based on FPV Installations Deployed on Pit Lakes in Former Mining Sites by 2050 in Poland

Mateusz Sikora [1,*] and Dominik Kochanowski [2]

1 Faculty of Civil Engineering and Resources Management, AGH University of Krakow, 30-059 Krakow, Poland
2 Sun Energy Group, 88-200 Radziejow, Poland
* Correspondence: sikoram@agh.edu.pl

Abstract: Green hydrogen production is expected to play a major role in the context of the shift towards sustainable energy stipulated in the Fit for 55 package. Green hydrogen and its derivatives have the capacity to act as effective energy storage vectors, while fuel cell-powered vehicles will foster net-zero emission mobility. This study evaluates the potential of green hydrogen production in Power-to-Gas (P2G) systems operated in former mining sites where sand and gravel aggregate has been extracted from lakes and rivers under wet conditions (below the water table). The potential of hydrogen production was assessed for the selected administrative unit in Poland, the West Pomerania province. Attention is given to the legal and organisational aspects of operating mining companies to identify the sites suitable for the installation of floating photovoltaic facilities by 2050. The method relies on the use of GIS tools, which utilise geospatial data to identify potential sites for investments. Basing on the geospatial model and considering technical and organisational constraints, the schedule was developed, showing the potential availability of the site over time. Knowing the surface area of the water reservoir, the installed power of the floating photovoltaic plant, and the production capacity of the power generation facility and electrolysers, the capacity of hydrogen production in the P2G system can be evaluated. It appears that by 2050 it should be feasible to produce green fuel in the P2G system to support a fleet of city buses for two of the largest urban agglomerations in the West Pomerania province. Simulations revealed that with a water coverage ratio increase and the planned growth of green hydrogen generation, it should be feasible to produce fuel for net-zero emission urban mobility systems to power 200 buses by 2030, 550 buses by 2040, and 900 buses by 2050 (for the bus models Maxi (40 seats) and Mega (60 seats)). The results of the research can significantly contribute to the development of projects focused on the production of green hydrogen in a decentralised system. The disclosure of potential and available locations over time can be compared with competitive solutions in terms of spatial planning, environmental and societal impact, and the economics of the undertaking.

Keywords: former mining sites; reclamation; renewable energy sources; floating photovoltaic plant; power-to-gas; green hydrogen; RES; FPV; P2G

1. Introduction

When the European Green Deal strategy was adopted in 2019, UE activities and projects aimed at protecting the climate became more intensive and accelerated, resulting in the proposal of the Fit for 55 package, adopted in July 2021 [1–3]. The main objective was to revise and adapt the EU climate law to meet the new objective of a 55% reduction in CO_2 emissions by 2030 and achieving climate neutrality by 2050 [4–7].

The EU Commission emphasised that in order to meet these targets, major technological changes are required both in power engineering systems and in other industries, such as transportation, construction, or agriculture [8–10]. These objectives are included in strategic documents adopted in Poland, and the Polish Energy Policy by 2040 (PEP2040) stipulates that the share of renewable energy sources in the gross energy consumption should be 23% by 2030 and the greenhouse gas emissions in relation to 1990 emission levels should be reduced by 30%. At the same time, the proportion of coal used by the power sector should be constantly reduced. In this context, green hydrogen production, regarded as the fuel of the future, may be one of the strategies with which to meet the targets specified in the European Green Deal [11,12]. Furthermore, hydrogen can be effectively used for energy storage, integrated with the power grid, enhancing its stability [13,14]. To date, the intermittent availability of energy from renewables (wind power and solar systems) has limited the increase of its share in the structure of national power grids [15]. The key postulate in the PEP2040 regulation is the development of low-emission transport, with the ultimate goal being zero-emission urban mobility systems in cities with more than 100,000 inhabitants by 2030. These objectives and regulations contained in the Sustainable Transport Development Strategy by 2030 are in line with this policy, stipulating that the urban mobility in Poland should rely on novel propulsion solutions, including electric cars or vehicles powered by alternative fuels, such as LNG, CNG, ammonia, or hydrogen cells.

According to the EU strategy supporting hydrogen technology, large-scale production of hydrogen should be effected in electrolyser systems with the capacity of up to several hundred megawatts (starting from 100 MW) [16]. In the variant adopted in Poland, the electrolysers are to be powered with surplus capacities from offshore wind farms [17,18]. This strategy might cause the overloading of high-voltage lines, particularly if the projected nuclear power plant is to be located in the seaside village of Lubiatowo-Kopalino, within the municipality of Choczewo.

The design of electrolysers for operation in adverse marine conditions was undertaken by the consortium of Orsted, Siemens, Gamesa Renewable Energy, Element Energy, and ITM Power. This project marks the first step in the development of a commercial-scale method for hydrogen production from offshore wind farms, while the hydrogen is to be generated offshore and transported via a pipeline. This solution ought to foster commercial implementations of hydrogen production systems from offshore wind farms, marking an interesting trend in the development of the hydrogen technology market in Poland. However, the main barrier faced by large-scale producers of green hydrogen from offshore wind farms lies with high transport costs if the dedicated pipeline system has yet to be constructed and with the technology of hydrogen production from marine water, which is still commercially unviable.

There is a certain niche, i.e., the potential of decentralised green hydrogen production in small Power-to-Gas systems [19,20]. Electric power can be converted into gas and stored as reserve for future use (i.e., grid balancing), further distributed, consumed by end-users, or further processed [21,22]. This potential is offered by pit lakes left over after the extraction of minerals (natural aggregates, sand) under wet conditions. Under Polish law, the mining sites need to be reclaimed once the mining operations cease. Due to their previous industrial history, water reservoirs in those sites offer an attractive location for floating photovoltaic facilities and P2G systems generating green hydrogen via electrolysis of H_2O.

The topic of developing post-mining areas for the purposes of electricity production and storage has already been discussed. A combination for the development of post-mining areas in the form of electricity production using FPV and energy storage has been proposed by Mert and Dincer. A unique integration of floating photovoltaic with underground energy storage and hydrogen energy storage systems, as well as a heat pump-driven district energy system, is analysed using a thermodynamic approach from energy and exergy points of view. The proposed design exploits unutilised natural bodies and abandoned structures in a sustainable manner [23]. Temiz and Javani describe another project that investigated the possibility of floating PV plant integration with existing basins for wastewater treatment. According to their calculations, water between 15,000 and 25,000 m^3 is saved by each MWp. Furthermore, there is a yearly energy yield of up to 10% through the cooling effect [24].

The sources of hydrogen production from renewables can potentially be engaged in providing energy flexibility in the context of involving customers, implementing novel solutions (such as electrolysers), and supporting middle- and low-voltage systems at the local level [25]. It is assumed that the flexibility services should be traded on energy flexibility platforms, and the remuneration thereof can be regarded as an alternative benefit of green hydrogen generation. In the event of an electricity surplus, the hydrogen generation should commence, while in the case of a power shortage, the hydrogen generation should be discontinued and all electricity should be delivered to the grid to meet the current demand. Obviously, these operations need to be synchronised, economic aspects duly taken into account, and business models developed for the key stakeholders.

It has been reported in the literature that the costs of the power supply to the electrolysers must include the high distribution fee. This conclusion could be considered premature, and, in the more secure variant, the electrolyser modules should be implemented in the direct vicinity or close to energy sources [26,27]. Taking into account this uncertainty and the risks involved, it is reasonable to suppose that decentralised, small-scale installations should act as a safety buffer in green hydrogen generation at the national level.

Development work and feasibility studies are being carried out around the world. Publication [28] shows that a Power-to-Gas installation based on FPV can achieve IRR rates of return of 20% in Indonesian conditions.

With regard to the location of P2G facilities, they do not require extensive networks of power transmission infrastructure, thus eliminating the need to obtain grid connection permits from distribution system operators [29,30]. It might appear therefore that there should be no restrictions as to their potential locations. Yet, the expertise gained when implementing projects involving electricity generation from renewables clearly indicates that before a stand-alone wind, solar, or water power facility can be built, numerous constraints have to be met, mostly in terms of environmental and social impacts. It is reasonable, therefore, to expect similar difficulties when implementing P2G systems.

When floating photovoltaic facilities located in lakes in former mining sites are integrated with electrolysers or the systems of electrolysers, the abovementioned administrative and legal obstacles (i.e., restrictions imposed by local land development plans or environmental and social concerns) should be removed. Once the mineral extraction operations under wet conditions (categorised as projects producing severe environmental impacts) receive the required approvals relating to land development and environmental and social impacts, it is reasonable to suppose that after the mining operations ceased, the respective administrative decisions could be easily revised because the environmental impacts of photovoltaic facilities are less severe and risks are lower than in the case of new localities. FPV installations have several advantages. The reuse of degraded areas, mitigating land-use conflicts, more efficiency than ground-mounted PV solar, saving water, improving resilience, and diversifying electricity-producing infrastructures are of FPVs key benefits [31].

There are abandoned mine lands nationwide where minerals used to be extracted under dry conditions (above the groundwater level), which can potentially be reclaimed and reused for industrial purposes (electricity generation from renewables or green hydrogen production) [32,33]. However, these lands are often attractive for reclamation and reuse as farming land or forests, reflecting the preferences of local communities [34]. These aspects are not addressed in this study. As they are located in areas degraded by prior mining activity, pit lakes are often remediated, providing space for recreation, yet such forms of redevelopment are suitable only in the vicinity of major cities, where recreation spaces are readily accepted if not requested by inhabitants. In large cities, there is still demand for rehabilitated water reservoirs to be reused for recreation; the most popular are those located in close proximity to city centres. Good examples of such reservoirs include Zakrzowek Lake and Bagry Lake in Cracow, the open-air bathing spot Morskie Oko in Wrocław, Zwirownia Lake in Reszów, Tarnobrzeskie Lake in Tarnobrzeg, Pogoria I and Pogoria II Lakes in Dabrowa Górnicza, Balaton Lake in Trzebinia, and the Kamionka Bolko reservoir in Opole. Considering the multitude of man-made lakes formed during mineral extraction operations under wet conditions, only a few of them are rehabilitated and reused for recreation.

Obviously, pit lakes are mostly located outside city centres and in areas where aggregate reserves were proved and administrative decisions could be issued to start the enterprise; in most cases, they are some distance from densely populated areas.

The key advantages of accommodating photovoltaic facilities on water reservoirs over locations on land include the higher energy efficiency due to the lower temperature of floating modules and the fact that it should not interfere with the existing infrastructure on land [35].

The key aspect in promoting the implementation of P2G systems relying on electric power supply from floating photovoltaic plants is the ready access to process water. Apart from other obvious advantages, this feature renders the P2G systems an attractive option for investors. The cost of water access is a major component of the variable operating costs; the average demand for water for electrolysers varies from 9 to 11 l/kgH_2 [36].

This paper contributes significantly to the scientific community by exploring the potential of green hydrogen production through Power-to-Gas systems integrated with FPV installations on pit lakes in former mining sites in Poland. This research is pioneering in evaluating the feasibility of using rehabilitated mining sites for renewable energy production and hydrogen generation, which is crucial for achieving climate neutrality by 2050. The study uses advanced GIS tools to identify potential sites for FPV installations, assess their capacity for hydrogen production, and model the timeline for site availability, offering a novel approach to decentralising green hydrogen production.

The findings can inform future projects focused on renewable energy and hydrogen production, providing a model for assessing similar sites globally, especially in the Member States of the European Union. By addressing the technical, legal, and organisational challenges associated with such installations, the paper adds valuable insights into the sustainable redevelopment of former industrial areas, contributing to broader efforts in climate change mitigation and sustainable energy transitions.

The paper addresses the following key research questions:

- Is there potential for green hydrogen production in Power-to-Gas systems operated on pit lakes in former mining sites in Poland by 2050?
- How can GIS tools be used to identify and evaluate suitable locations for floating photovoltaic installations on these pit lakes?
- What are the technical, organisational, and legal challenges associated with implementing FPV-P2G systems on rehabilitated mining sites?
- What is the projected capacity of green hydrogen production from these systems, and how can it support urban mobility in terms of powering city buses in Poland's West Pomerania province by 2030, 2040, and 2050?

The answers to these questions form the basis for a comprehensive assessment of the potential for green hydrogen production in the study, providing an outline of a roadmap for future developments in this area. To deepen knowledge and refine the direction of this development, further research should be conducted, particularly focussing on the economics of the proposed venture, the logistics of transporting hydrogen supplies to urban agglomerations, and a detailed environmental impact assessment.

2. Method of Assessing the Potential of Floating Photovoltaic Plant Implementation on Pit Lakes

The method of assessing the potential of accommodating floating photovoltaic plants on pit lakes being rehabilitated is universal. Thus, when all geospatial data required for the analysis are available, the method can be effectively employed to examine each potential location and site-specific boundary conditions.

The key element of the algorithm is the database of national geospatial data BDOT10k made available in electronic form via the browser service, in accordance with the Regulation of 27 July 2021 by the Minister of Development, Labour, and Technology, having relevance to datasets of topographic objects, general geographic objects, and standard cartography studies. This Regulation is the executive act of the Law on Geodesy and Cartography of 17 May 1989 (Journal of Laws 2023, item 1752, consolidated act of 31 August 2023). The BDOT10k dataset and the data made available as public information by the Polish Geological Institute-National Research Institute are the input for geoprocessing analysis, yielding the sequence of intermediate results.

Basically, the algorithm involves three stages:

- Stage I—definition of analysed regions.
- Stage II—quest for available locations for FPV plants.
- Stage III—scheduling the availability of areas for investment.

The structure and operating principle of the algorithm are shown as a block diagram in Figure 1.

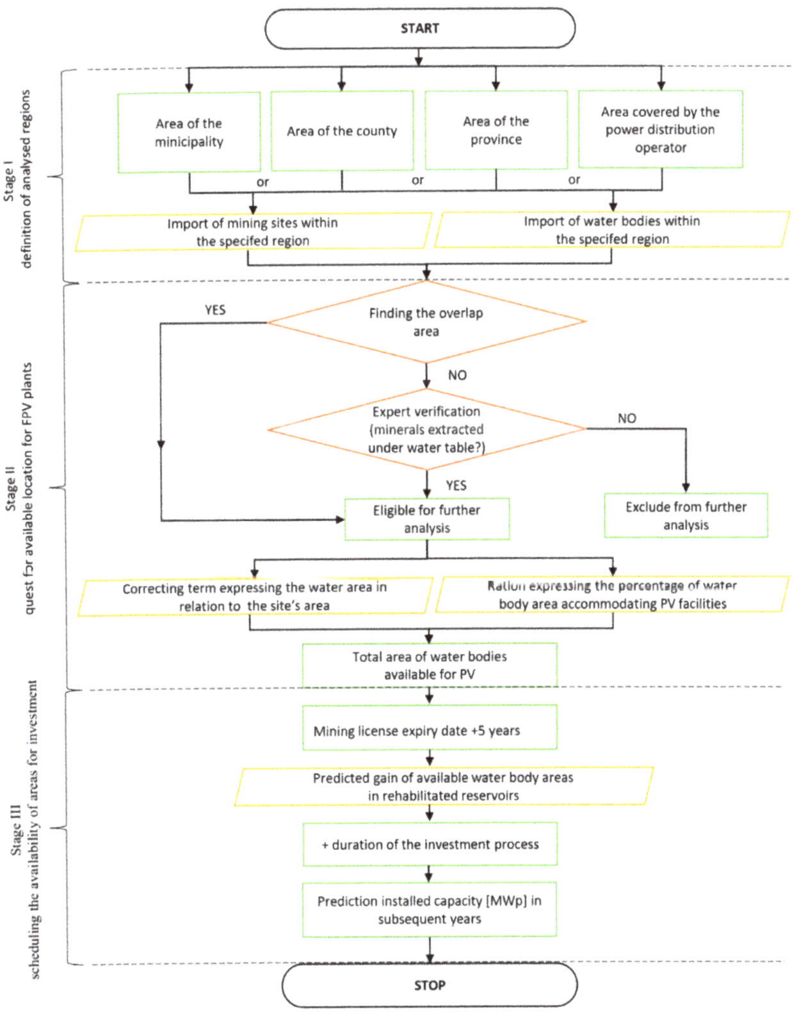

Figure 1. Algorithm in the universal method of assessing the potential of implementing FPV facilities on rehabilitated pit lakes in former mining sites (author's own sources).

2.1. Stage 1—Definition of the Analysed Regions

In Stage 1, the site boundaries are to be defined for the algorithm to yield the result. Considering the fact that the results are of geostrategic importance, it is reasonable to expect that the analysed area should be either the entire area within state borders, a formally defined administrative unit, such as a province, county, or municipality, or other areas with formal or administrative boundaries, such as the areas covered by the electricity distribution operators. The area explored in the analysed case study is the West Pomerania province, and the geospatial data are provided by respective geodesy offices. In the case of Poland, the main sources are data that include the database of geographic features BDOT10k, particularly with respect to land features, listed in Article 3, Section 4.

Land survey data are complemented with data made available by the Polish Geological Institute-National Research Institute, relating to locations of mineral deposits and active mining areas subject to mining license under the Geological and Mining Law of 9th June 2011 (Journal of Laws 2023 item 633, consolidated act of 3 April 2023).

2.2. Stage II—Quest for Locations Available for FPV

In Stage II, the geoprocessing tools are deployed to locate the mineral deposits that may extend beyond the analysed site and to determine the category of the object OT_PTWP_A—surface water with the attribute value 03—stagnant water. All areas found in this stage are subjected to further analysis. Considering the insufficient speed of BDOT10k data updating and dynamic changes of surface conditions in surface mining, the remaining (excluded) regions should be verified manually in relation to the current orthophotomap available in geoportals (for Poland), and, in the case of a large number of areas considered, the analysis should be supported by visual image identification tools. The latest available satellite imagery is included in the set of rejected mining areas. Several dozen areas are visually assessed to confirm whether they have been correctly excluded from the calculation procedure or if they are located on stagnant water bodies not yet entered into the BDOT10k database. This solution can be automated in the future. Selected areas are then subject to the standard procedure, while the remaining ones categorised as surface mining sites where minerals are to be extracted above the groundwater table level are thus excluded.

To determine the target areas for accommodating floating photovoltaic plants, it is required that the status of active mining operations be first established. In the case of sand or gravel extraction under the water table, the widely adopted rule is that the documented deposit lies within the limits of the business property remaining in the hands of the entrepreneur and that the active mining site is determined in accordance with the plot boundary plans. The target excavation is profiled in accordance with the current standards, retaining the required safety margins to separate it from real estate that is not under the management of the entrepreneur. It is reasonable to suppose, therefore, that the deposit is extracted by full contour mining or within the contour limited by the boundaries of the mining area, leaving the non-production margins as stipulated by the law.

The sketch in Figure 2 shows the spatial orientation and two excavations in different stages of progressing mining operations.

In excavation 1, the extraction operations have ceased, and the water reservoir has its ultimate contour. The surface area of the documented mining site is 2.425 ha, and the surface area of the water reservoir is 2.087 ha. The area occupied by the water reservoir accounts for 86% of the total area to be surface mined.

Applying the same rule to excavation 2, it is reasonable to expect that the excavation should reach its final profile in the following years (as of or prior to the mining concession expiry date) and the water reservoir should have a surface area of 5.964 ha, while the area of the mining site is 7.190 ha. The water reservoir accounts for 83% of the site area.

The analysis of projects implemented in this area seems to confirm the adequacy of the adopted rule. In the calculation procedure, the proportion of the water table surface area in the reclaimed area is taken to be 80%, which is a pessimistic result in relation to the analysed case studies in the West Pomerania province. This result seems adequate and readily implementable without repeating the verification for mineral deposits in lowland areas, while for areas with diverse morphologies (mostly found in the south of Poland), the procedure needs to be repeated and the percentage of the water body area in relation to the site's surface area should be confirmed or recalculated. In morphologically diverse terrain, a lower proportion of water surface area should be expected within the mining area boundaries due to the profiling of open pits, which are partially non-waterlogged.

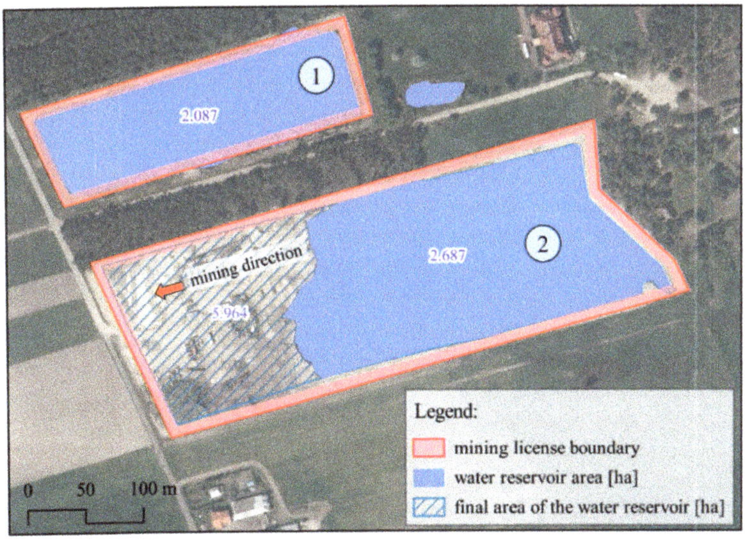

Figure 2. Predicted area of water reservoirs based on identification of proved mineral reserves and in the light of standing mining concessions (author's own sources).

The other constraint limiting the scope of FPV investments is the proportion of the water reservoir area that is available for the installation of FPV facilities; the major determinants are the actual shape of the reservoir and irregular shoreline.

Of particular importance are site management restrictions precluding its reuse for industrial purposes, prompting other reclamation schemes. It could be postulated that all available areas should be effectively utilised and, when making strategic decisions, the assessment of potential green hydrogen generation should rely on the highest possible values of the water coverage ratio. On the other hand, it is unreasonable to expect that all available pit lakes in former mining sites should be put to such use. Reusing a fraction of water bodies for the deployment of FPV is fully in line with the principles of sustainable development. In practical terms, there are no obstacles precluding the use of the part of the lake not covered with FPV modules for other purposes, for industry, for natural site recovery, or, under specified circumstances, also for recreation [37,38].

In the recalled case study, the underlying assumption was that 50% of the available water surface area should be covered by floating FPV modules. Ultimately, that means the coverage of half of the water surface, or, in particularly suitable locations, the coverage ratio could be 100%, with further investments. Actually, a coverage ratio of 50% should be regarded as the average value for the analysed region (West Pomerania province).

Figures 3 and 4 refer to two projected photovoltaic installations covering approximately 50% of the available water surface area. In the case illustrated in Figure 3, the southern part of the lake is left uncovered and can be put to different use; in the case shown in Figure 4, the photovoltaic modules should cover the majority of the available water surface due to irregular shorelines.

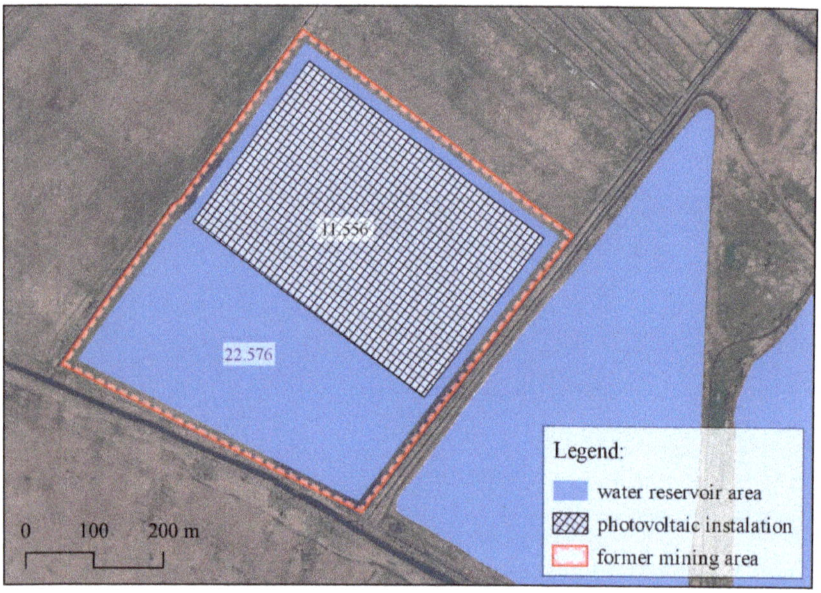

Figure 3. 50% coverage with FPV modules for the water reservoir with regular shoreline (author's own sources).

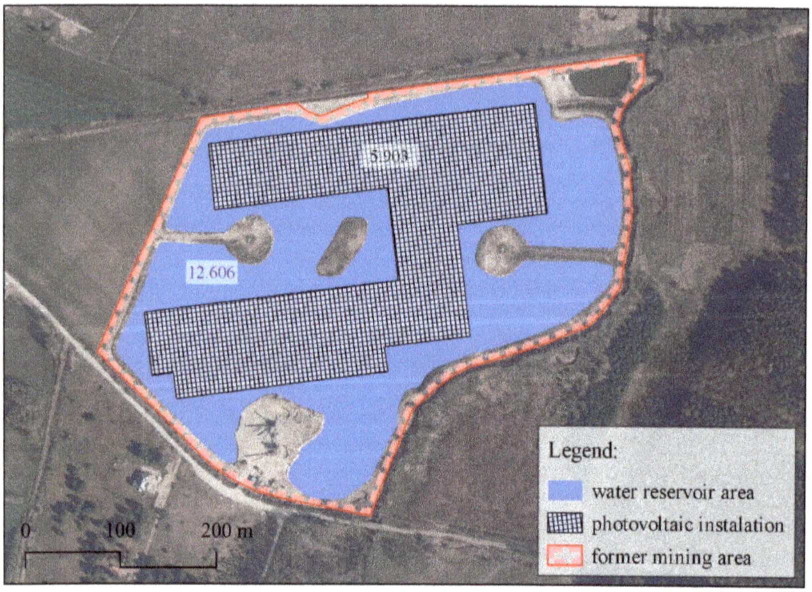

Figure 4. 50% coverage with FPV modules for the water reservoir with irregular shoreline (author's own sources).

In the final step of Stage II, the total area available for FPV deployment should be determined.

2.3. Stage III—Scheduling the Availability of the Area for Investment

In the final stage, the sites selected for investment should be arranged in an order demonstrating their ready availability in time, showing the increment of available areas. The scale of this increment depends on the provisions of the mining law and relevant regulations governing the implementation of P2G installations incorporating floating photovoltaic modules and an electrolyser or a set of electrolysers:

- Geological and Mining Law of 9 June 2011 (Journal of Laws 2023.63, consolidated act of 3 April 2023)
- Law on Protection of Agricultural and Forest Land of 3 February 1995 (Journal of Laws. 2024.82, consolidated act of 24 January 2024
- Law on Public Access to Environmental Information, Public Participation in Environment Protection, and Environmental Impact Assessment of 3 October 2008 (Journal of Laws 2023.1094, consolidated act of 12 June 2023)
- The Energy Law of 10 April 1997 (Journal of Laws 2024.266, consolidated act of 28 February 2024)
- The Water Law Act dated 20 July 2017 (Journal of Laws 2023.1478, consolidated act of 1 August 2023)
- The Construction Law dated 7 July 1994 (Journal of Laws 2024.725, consolidated act of 14 May 2024)

Former mining sites are to be rehabilitated when the lands (and water bodies) are wholly or partly abandoned or left unused for a specified period of time. The reclamation project must be completed by the mining company no later than 5 years after the land has been declared useless. In the analysed case study, the maximum allowable waiting period is taken, so it is assumed that the pit lake should be rehabilitated within 5 years from the expiry of the valid mining license. In practical terms, mining concessions can be issued for periods longer than those based on the current progress of mining operations and the quantity of recoverable reserves. In such cases, land reclamation schemes will be implemented sooner, the allowable time period starting from the moment when the deposit is exhausted instead of the expiry date of the mining concession. Furthermore, mining sites can be rehabilitated in part while the mining license is still standing. Hence, the assumption adopted for the algorithm serves as the worst possible scenario, and the available surface area of water bodies in the analysed region may increase sooner than predicted by the algorithm.

The ceasing of mining activities and converting the site to be rehabilitated for use as a water reservoir is the prerequisite to initiating the investment project aimed at P2G installation. Approximately, it takes about two years to obtain the decision on environmental conditions and on water management, as well as building permits for investment projects in abandoned mining areas. In consideration of the legal status of these sites and the activity of other operators, these projects ought to not evoke social or environmental concerns. Poland to date has had no experience in the construction of P2G installations relying on floating photovoltaic plants and electrolysers, yet the entire investment process and hence the time to obtain the requisite administrative decisions and to initiate engineering work is nearly the same as when planning a standard photovoltaic plant with energy storage facilities. The construction stage usually takes a few months; larger installations may take over a year to complete. In the simulated variant, the P2G installation is to be deployed for industrial use after 7 years from the expiry of the mining license.

Strategic plans for floating photovoltaic plants typically provide the installed capacity of 1 MWp per 1 ha of the water surface area, allowing the projected capacity and timeframe of implemented photovoltaic farms to be effectively scheduled. In the analysed case, the project scheduled is the P2G system to be implemented in the West Pomerania province by 2050.

3. Potential Effects of P2G System Deployment on Pit Lakes in Former Mining Sites

In the case of photovoltaic plants, power output predictions depend on the solar irradiance factor. Numerous authors have reported that the installation of floating photovoltaic modules is beneficial for the FPV plant's performance. Güllü, E., et al. described that lab-made modified graphite cathodes were used to design and implement floating PV-assisted alkaline electrolysis cells. The results showed that floating PV panels were more beneficial than land-mounted panels, and the G/Ni/Co enhanced the hydrogen generation performance of the system [39]. Due to the water-cooling effect, the operating cell temperature is significantly lower than the temperature of modules installed on land. This lower temperature can be attributed to the module being accommodated on the water surface and to the flow of circulating air, which tends to be cooler in the neighbourhood of water reservoirs than on land [40,41].

The operating cell temperature is the key determinant of each PV plant's efficiency. The well-known formula presented by Kamuyu (Kamuyu et al., 2018) [42] and recalled in this study takes into account the water-cooling effect to obtain the cell operating temperature in FPV modules.

$$T_C = e_0 + e_1 \cdot T_a + e_2 \cdot G_T - e_3 \cdot V_w \tag{1}$$

where T_a—air temperature, G_T—solar radiation, V_w—wind speed, e_0 = 2.0458 °C, e_1 = 0.9458 °C^{-1}, e_2 = 0.0215 °C·m^2·day·kWh^{-1}, and e_3 = 1.2376 °C·s·m^{-1}, are empirically determined coefficients.

At the latitude of Poland and in average atmospheric conditions thereof, yearly electricity generation in this type of plant is up to 1100 MWh per 1 MWp. This value is assumed in the developed model, while the algorithm will admit any value predicted in consideration of specific local settings and average weather conditions, such as the solar zenith angle, the type of PV modules, and the operating time. Table 1 summarises the simulated predicted electricity generation in floating photovoltaic plants accommodated on rehabilitated pit lakes in former mining sites by 2050.

Table 1. Predicted electricity generation in FPV installations deployed on pit lakes in former mining sites.

Year	Increment of Lake Surface Area [ha]	Increment of Lake Surface Area Available for FPV Installation [ha]	Increment of Installed Capacity [MWp]	Predicted Electricity Generation [kWh]
2024	22.73	11.36	-	-
2025	7.28	3.64	-	-
2026	61.44	30.72	11.36	12,501,324
2027	17.65	8.82	15.00	16,504,620
2028	12.14	6.07	45.73	50,298,556
2029	10.37	5.19	54.55	60,004,604
2030	109.82	54.91	60.62	66,681,296
2031	19.66	9.83	65.81	72,385,764
2032	10.40	5.20	120.71	132,784,168
2033	2.58	1.29	130.54	143,598,928
2034	2.56	1.28	135.74	149,318,004
2035	32.55	16.27	137.03	150,735,508
2036	23.25	11.63	138.31	152,140,780
2037	-	-	154.58	170,042,972
2038	-	-	166.21	182,831,352
2039	51.84	25.92	166.21	182,831,352
2040	9.67	4.83	166.21	182,831,352
2041	19.03	9.52	192.13	211,342,296
2042	-	-	196.96	216,659,960
2043	0.75	0.38	206.48	227,127,384

Table 1. *Cont.*

Year	Increment of Lake Surface Area [ha]	Increment of Lake Surface Area Available for FPV Installation [ha]	Increment of Installed Capacity [MWp]	Predicted Electricity Generation [kWh]
2044	-	-	206.48	227,127,384
2045	54.16	27.08	206.86	227,541,336
2046	13.40	6.70	206.86	227,541,336
2047	2.46	1.23	233.94	257,330,480
2048	-	-	240.64	264,703,076
2049	-	-	241.87	266,054,932
2050	51.65	25.82	241.87	266,054,932

The estimation in Table 1 results from applying the model proposed in Figure 1. The forecast of potential electricity production takes into account the generation of 1100 MWh per 1 MWp of the installed power capacity of the power plant. This value is commonly adopted in feasibility studies for photovoltaic power plants in Poland.

The average annual irradiance value was adopted according to the PVGIS-SARAH2 database. For the West Pomeranian Voivodeship, the value was estimated to be 1315 kWh/m^2. Depending on other variables affecting the efficiency of electricity production (system loss, PV mounting position, slope angle, PV technology), this results in a yearly PV energy production level of 1100 MWh per 1 MWp of installed capacity.

Figure 5 illustrates potential rates of increase in electricity generation, showing annual growth (for new floating FPV plants).

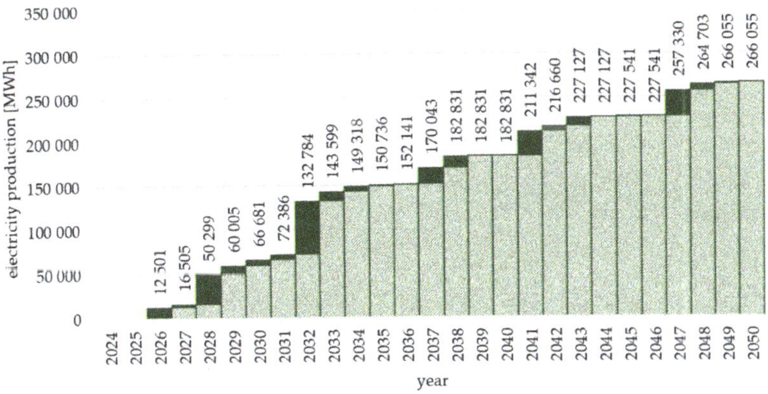

Figure 5. Predicted electricity generation in FPV facilities deployed on the pit lakes in former mine sites by 2050.

The integration of floating photovoltaic plants with an electrolyser or a set of electrolysers in the P2G system has numerous benefits. First of all, the process of green hydrogen production requires large amounts of process water and is energy-consuming [43–46]. Accommodating P2G installations and photovoltaic plants on a water reservoir provides stable access to process water. Producing green hydrogen at the locality of a renewable energy source enables the off-grid operation of the generating unit, i.e., without obtaining the grid connection permit. Currently, this is one of the major obstacles in the development of distributed renewable energy sources.

Poland is a manufacturer of modern hydrogen-fuelled city buses, which have been extensively tested and commissioned for use in Polish and European cities and also on other continents [47–52]. Field tests conducted in city agglomerations in Krakow and Vienna reveal that the city bus Mega (Solaris Urbino 18 Hydrogen) bus uses 20 kgH_2O/100 km, while the model Maxi (Solaris Urbino 12 Hydrogen) uses 6.3 kg on average [53]. Table 2 summarises the achievable levels of green hydrogen generation in facilities deployed on pit lakes in former mining sites, related to the size of the bus fleet and the annual fuel demand. According to the Central Statistical Office reports, the average value of km driven by a city bus in Poland is about 80,000 km per year. Generally, municipal transport companies distribute the size of the bus fleet in roughly equal proportions by type. Figure 6 shows the achievable increase in the number of city buses fuelled with hydrogen produced in distributed P2G systems in the West Pomerania province over the years 2030, 2040, and 2050.

Table 2. Achievable levels of green hydrogen generation related to consumption by net-zero emission urban mobility systems (MEGA and MAXI bus types).

Year	Predicted Electricity Generation [kWh]	Electricity Consumption Ratio [kWh/kgH_2]	Hydrogen Production [kgH_2]	Bus Range [km/year]	Bus Fleet Size-Number of Mega Busses [Item]	Bus Fleet Size-NUMBER of Maxi Buses [item]
2024	-	56.0	-	-	-	-
2025	-	56.0	-	-	-	-
2026	12,501,324	56.0	223,238	2,232,379	18	16
2027	16,504,620	56.0	294,725	2,947,254	23	22
2028	50,298,556	55.5	906,280	9,062,803	71	67
2029	60,004,604	55.0	1,090,993	10,909,928	86	80
2030	66,681,296	54.5	1,223,510	12,235,100	96	90
2031	72,385,764	54.0	1,340,477	13,404,771	106	98
2032	132,784,168	53.5	2,481,947	24,819,471	195	182
2033	143,598 928	53.0	2,709,414	27,094,137	213	199
2034	149,318,004	52.5	2,844,152	28,441,525	224	209
2035	150,735,508	52.0	2,898,760	28,987,598	228	213
2036	152,140,780	51.5	2,954,190	29,541,899	233	217
2037	170,042,972	51.0	3,334,176	33,341,759	263	245
2038	182,831,352	50.5	3,620,423	36,204,228	285	266
2039	182,831,352	50.0	3,656,627	36,566,270	288	268
2040	182,831,352	49.5	3,693,563	36,935,627	291	271
2041	211,342,296	49.0	4,313,108	43,131,081	340	317
2042	216,659,960	48.5	4,467,216	44,672,157	352	328
2043	227,127,384	48.0	4,731,821	47,318,205	373	347
2044	227,127,384	47.5	4,781,629	47,816,291	377	351
2045	227,541,336	47.0	4,841,305	48,413,050	381	355
2046	227,541,336	46.5	4,893,362	48,933,621	385	359
2047	257,330,480	46.0	5,594,141	55,941,409	441	411
2048	264,703,076	45.5	5,817,650	58,176,500	458	427
2049	266,054,932	45.0	5,912,332	59,123,318	466	434
2050	266,054,932	44.5	5,978,763	59,787,625	471	439

Simulations revealed that with water coverage ratio increase and with the planned growth of green hydrogen generation, it should be feasible to produce fuel for net-zero emission urban mobility systems to power 200 buses by 2030, 550 buses by 2040, and 900 buses by 2050 (for city bus models Maxi (40 seats) and Mega (60 seats)).

The Energy Policy of Poland until 2040 focusses on the development of low-emission transportation, with particular emphasis on achieving zero-emission public mobility by 2030 in cities with a population over 100,000. As of the end of 2022, the bus fleet of the West Pomerania Voivodeship consists of 534 units. This suggests that the estimated potential for green hydrogen production could provide approximately 35% of the zero-emission public mobility needs by 2030 and nearly 100% by 2040 throughout the voivodeship, not just in agglomerations with populations over 100,000.

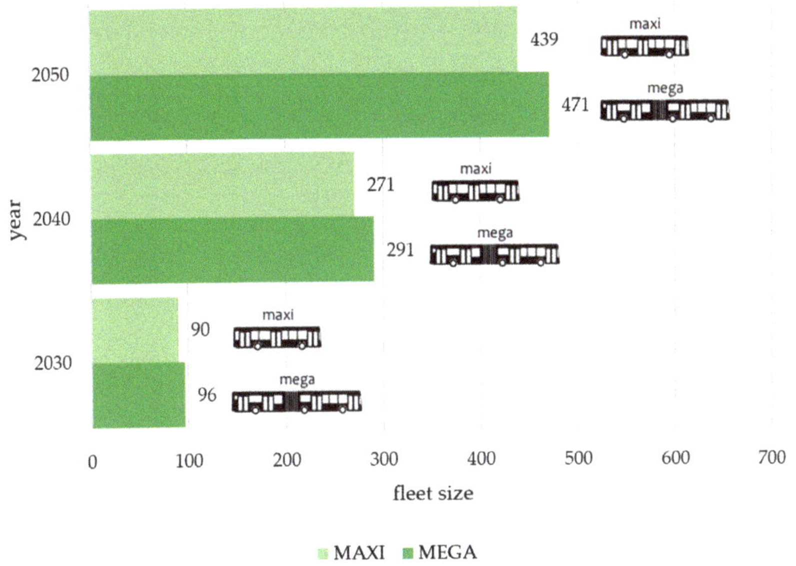

Figure 6. Predicted number of hydrogen-fuelled city buses powered from distributed P2G sources.

Data from the Central Statistical Office do not differentiate between MAXI and MEGA bus units. Therefore, any deviation from the proposed MAXI/MEGA unit ratio would result in a change in the per-unit hydrogen consumption.

4. Discussion

The article identifies new investment areas that have the potential to support zero-emission urban mobility initiatives. Current changes in spatial planning legislation in Poland are making it increasingly challenging to build photovoltaic plants on land. Nevertheless, such investment areas will still be available.

The performance of the developed algorithm was tested on an example of a case study of the West Pomerania province, leading to definite results and conclusions. Sand and gravel aggregate is extracted country-wide, as it is a vital component in the supply chain for the construction industry [54]. Due to the low value of the product, the key component of its market price is the transportation cost [55]. For those reasons, the proved reserves of sand and gravel aggregate are fairly uniformly distributed all over the country. According to the Balance of Mineral Resources Deposits as of 31 December 2022 (last published), there are 3674 sand and gravel deposits being developed in Poland. In some cases, sand and gravel are extracted under wet conditions (below the water table), leaving a legacy in the form of rehabilitated water ponds once the mining activities cease. These are not the only water reservoirs that emerge during surface mining activities. There are some other minerals that are extracted below the underground water table as well; alternatively, ponds can be formed as interior basins in impermeable strata.

With the development of road investment and construction projects, new resources of sand and gravel aggregates are explored and proven, new surface mining sites emerge, and, when the extraction ceases, are subject to rehabilitation and reclamation schemes, including the rehabilitation of water bodies [56].

The developed and tested model demonstrated that the West Pomerania province has the potential for green hydrogen production on post-mining reservoirs of sand and gravel deposits. The estimation of this potential was made possible through the use of GIS algorithms utilising data from databases provided by administrative bodies. The calculation and scheduling of production potential required the establishment of certain boundary conditions resulting from technical and organisational constraints, such as the shape of the reservoirs, the duration of the mining license, the period of technical reclamation of mining sites, and the legal and formal conditions for venture implementation.

The analysed case study of the West Pomeranian province showed that the potential for green hydrogen production in Power to Gas installations in off-grid technologies can be built on selected water reservoirs in former mining sites, thus solving the problems caused by limited transmission capacity of distribution networks. Floating photovoltaic power plants integrated with a set of electrolysers can be a source of green hydrogen for the region in a distributed system and be used to fuel zero-emission urban mobility based on fuel—cell buses. The demonstrated values of hydrogen production potential for the years 2030, 2040, and 2050 require further work and evaluation regarding their economic feasibility, as well as distribution and utilisation possibilities to achieve the goal of net-zero emission mobility.

Potential for Green Hydrogen Production in Power-to-Gas Systems Operated on Pit Lakes in Former Mining Sites in Poland by 2050

According to the obtained modelling results, it is reasonable to expect a steady growth of installed capacity and yearly electricity generation from FPV to 266 GWh in 2050, while its half value can be achieved in 2032. The service life of photovoltaic installation is about 30 years, yet considering the advancements in PV technology and development of upgraded PV modules with improved energy efficiency, they are likely to be replaced with newer technologies before their 30-year life is over. Thus, the actual electricity production levels can be higher than now predicted.

In fact, without extending and developing the national power transmission and distribution network, meeting the objectives specified in the Fit for 55 package and reaching climate neutrality by 2050 might be hard to achieve [57,58].

Green hydrogen generation requires process water supply to the electrolysers. The unit demand for process water varies from 9 to 11 $l(H_2O)/kg(H_2)$. In the calculations, a higher value of 11 $l(H_2O)/kg(H_2)$ was used. It means that in the case of a water body with a surface area of 2.0 ha (1.0 ha being covered with FPV modules, installed capacity of ~1 MWp) the amount of process water to be drawn is 216,071 L per year. Average annual precipitation in the West Pomerania region over the last few years has remained on the level of 500 mm (below that the country's average), which indicates that the annual demand for process water should not exceed 2% of precipitation supplying the water reservoir. It appears that drawing process water from the reservoir should not directly impact the water balance in the reservoir, while at the same time ensuring the supply of 100% process water.

Fluctuations of water level in the reservoirs are attributable to natural hydrological cycles involving surface run-off, infiltration, supporting vegetation, groundwater, retention (including bioretention), and evaporation and transpiration processes. Covering the water surface with FPV modules will reduce evaporation, which indirectly impacts water retention in the reservoir. Actually, each project involving the implementation of P2G systems is subject to environmental impact assessment, also in terms of process water demand, yet according to general estimates, in most cases the negative impacts should be negligible.

In the light of new technological developments, it is reasonable to expect that energy demand for green hydrogen generation should be gradually reduced. The current demand is at a level of 55.6 kWh/kg H_2O and is expected to fall to 44.5 kWh/kg H_2O in 2030 [48]. This downward trend is expected to continue until 2050, though the estimates are based on the energy consumption level falling down to 44.5 kWh/kg H_2O, to rigorously adhere to underlying assumptions.

To evaluate the potential development of green hydrogen generation in distributed P2G facilities in the West Pomerania province, the study is conducted on the feasibility of using green hydrogen in fuel cells for urban mobility systems [59,60].

On account of a large number of small, distributed sources, hydrogen can be transported within the province by road and supplied to the urban distribution network; for example, for powering the urban mobility systems.

Due to the distributed production model with small volumes, on-site gas storage is not anticipated. The road infrastructure, connected to the current mining production areas, is adapted for the export of hundreds of thousands of tonnes of aggregate. Therefore, there will be no need to expand the road network for the regular transportation of hydrogen by truck to transportation companies in nearby urban agglomerations.

In the proposed model, green hydrogen production is scheduled such that it should enable the fleet of nearly 1,000 city buses to be powered by 2050. According to the West Pomerania bus inventory data as of 31 December 2022, provided by the Central Statistical Office, the fleet size was 534 units. This estimate relies on the number of Maxi and Mega buses, while urban mobility in smaller agglomerations is largely dependent on Midi vehicles, providing transportation for a minimum of eight people, in accordance with the Road Transport Law of 6 September 2001 (Journal of Laws 2024.728, consolidated act of 15 May 2024). These vehicles have lower fuel consumption; therefore, it is reasonable to suppose that green hydrogen supply to net zero-emission urban mobility systems should be higher than the estimated value.

The developed schedule shows that the West Pomeranian province has the potential to produce green hydrogen from distributed sources located on reclaimed water reservoirs in former mining sites. Furthermore, the model demonstrates that the demand for net zero-emission fuel for all buses deployed by urban mobility systems within the province should be fully met as early as 2038.

Attention must be given to geolocational variables that have a crucial impact on the estimation results. These include the quantity of mineral deposits, prevailing geological and mining conditions, exploitation methods, terrain morphology, and geographic latitude and longitude. The model's operating principle is universal and can be successfully applied in other regions of Poland. Furthermore, with access to geospatial data from other countries and adaptation of the model to the prevailing legislation, estimations can be made for other nations and regions. This is particularly feasible for EU member states, whose legal regulations are aligned with the prevailing Directives of the European Parliament, making the modification and application of the model relatively straightforward.

Despite the fact that, compared to primary extraction activities, the production of green energy and green hydrogen in post-exploitation areas objectively has a lesser impact on the natural environment, a significant barrier may be the reluctance of local communities to neighbour this new type of enterprise. For FPV projects combined with hydrogen production to have a chance of success, it is essential to invest in public education from the initial investment stage.

To confirm the economic viability of the proposed green hydrogen production for urban mobility systems, a series of modelling studies are necessary, including economic models, to compare them with alternative solutions such as Power-to-Gas (P2G) production based on land photovoltaic plants or wind power plants.

5. Conclusions

The study presented examines the potential for green hydrogen production using Power-to-Gas systems integrated with floating photovoltaic installations on former mining lakes in Poland, forecasting to 2050. This novel approach not only highlights the feasibility of utilising unused water bodies for renewable energy production, but also addresses critical energy and emission issues faced by regions undergoing energy transition.

The results show that integrating FPV systems with P2G technology at former mining sites presents a significant opportunity for green hydrogen production. The use of FPV in former mining lakes, often considered environmentally burdensome, transforms these sites into valuable assets in the renewable energy landscape. The modelling and analysis of the study suggest that the energy generated by FPV systems can be efficiently converted to hydrogen, which serves as a versatile energy carrier. This approach is in line with global efforts to decarbonise both energy and transport.

One of the key contributions of this study is the identification and quantification of the potential hydrogen production capacity in specific post-mining lakes in Poland. By integrating advanced geospatial data, the study provides a comprehensive assessment of the potential of FPV-P2G systems to contribute to the decarbonisation of public mobility. The results reveal that with appropriate technological and infrastructure investments, these systems can play a key role in the Polish hydrogen economy by 2050, offering a sustainable alternative to fossil fuels.

The study fills a significant gap in the literature by addressing the intersection of renewable energy production and post-mining land use. Previous studies have largely focused on land-based PV systems, often on large-scale agricultural areas, or have treated hydrogen production and PV plants as separate domains. This paper brings these areas together, offering a coherent strategy for integrating renewable energy generation with hydrogen production in a way that leverages existing land-use conditions. The novelty of this study lies in its interdisciplinary approach, combining the successive remediation of post-mining areas with renewable energy and hydrogen production.

Furthermore, this study provides valuable insights for researchers, policymakers, and stakeholders involved in the transition to zero-emission urban mobility. The results suggest that decentralised, small-scale FPV-P2G installations can serve as key elements of a national green hydrogen production strategy, especially in countries with similar post-industrial landscapes and legal regimes. The study also highlights the importance of addressing potential challenges, such as environmental impact and social acceptance, that are key to the successful implementation of these ventures.

The developed model and research results contribute to the growing body of knowledge on green hydrogen production and offer a scalable, innovative solution to post-mining area development challenges. The study not only addresses existing gaps in the literature but also lays the foundation for future research that could investigate the optimisation of FPV-P2G systems in post-mining areas, the integration of other renewable energy sources, and the socio-economic implications of widespread implementation. The results of this study have the potential to impact both research and practical applications, contributing to the global transformation towards a sustainable energy future.

Author Contributions: Conceptualization, M.S.; methodology, M.S. and D.K.; software, M.S.; validation, M.S.; formal analysis, M.S. and D.K.; investigation, M.S. and D.K.; resources, M.S.; data curation, M.S.; writing—original draft preparation, M.S.; writing—review and editing, M.S.; visualization, M.S.; supervision, D.K.; project administration, M.S. and D.K.; funding acquisition, M.S. and D.K. All authors have read and agreed to the published version of the manuscript.

Funding: This research was funded by AGH University of Krakow, Faculty of Civil Engineering and Resource Management; subsidy number: 16.16.100.215, 501.696.7996/L34.

Data Availability Statement: The data can be accessed upon request any of the corresponding author.

Conflicts of Interest: The authors declare no conflict of interest.

References

1. Soares, A.G. The European Green Deal. *Rev. Jurid. Portucalense* **2024**, *35*, 44–67. [CrossRef]
2. Schlacke, S.; Wentzien, H.; Thierjung, E.M.; Köster, M. Implementing the EU Climate Law via the 'Fit for 55' package. *Oxf. Open Energy* **2022**, *1*, oiab002. [CrossRef]
3. Wilson, A. *Revision of the Renewable Energy Directive: Fit for 55 Package*; European Parliament: Brussels, Belgium, 2021.
4. Ministry of Climate end Environment Republic of Poland. *The Energy Policy of Poland until 2040*; Annex to the Notice of the Minister of Climate and Environment Dated 2 March 2021 (item 264); Ministry of Climate end Environment Republic of Poland: Warsaw, Poland, 2021.
5. Bellora, C.; Fontagné, L. EU in search of a Carbon Border Adjustment Mechanism. *Energy Econ.* **2023**, *123*, 106673. [CrossRef]

6. Rasmussen, M.G.; Andresen, G.B.; Greiner, M. Storage and balancing synergies in a fully or highly renewable pan-European power system. *Energy Policy* **2012**, *51*, 642–651. [CrossRef]
7. Perissi, I.; Jones, A. Investigating European Union Decarbonization Strategies: Evaluating the Pathway to Carbon Neutrality by 2050. *Sustainability* **2022**, *14*, 4728. [CrossRef]
8. The New Industrial Strategy for Europe. *Intereconomics* **2021**, *56*, 132. [CrossRef]
9. Hetland, J.; Zheng, L.; Shisen, X. How polygeneration schemes may develop under an advanced clean fossil fuel strategy under a joint sino-European initiative. *Appl. Energy* **2009**, *86*, 219–229. [CrossRef]
10. Rabbi, M.F.; Popp, J.; Máté, D.; Kovács, S. Energy Security and Energy Transition to Achieve Carbon Neutrality. *Energies* **2022**, *15*, 8126. [CrossRef]
11. Lux, B.; Pfluger, B. A supply curve of electricity-based hydrogen in a decarbonized European energy system in 2050. *Appl. Energy* **2020**, *269*, 115011. [CrossRef]
12. Garcia, D.A.; Barbanera, F.; Cumo, F.; di Matteo, U.; Nastasi, B. Expert opinion analysis on renewable hydrogen storage systems potential in Europe. *Energies* **2016**, *9*, 963. [CrossRef]
13. Ciechanowska, M. *Hydrogen Strategy for a Climate-Neutral Europe*; Instytut Nafty i Gazu: Krakow, Poland, 2020. [CrossRef]
14. Berrada, A.; Laasmi, M.A. Technical-economic and socio-political assessment of hydrogen production from solar energy. *J. Energy Storage* **2021**, *44*, 103448. [CrossRef]
15. Calado, G.; Castro, R. Hydrogen Production from Offshore Wind Parks: Current Situation and Future Perspectives. *Appl. Sci.* **2021**, *11*, 5561. [CrossRef]
16. Wolf, A.; Zander, N. Green Hydrogen in Europe: Do Strategies Meet Expectations? *Intereconomics* **2021**, *56*, 316–323. [CrossRef]
17. Lux, B.; Deac, G.; Kiefer, C.P.; Kleinschmitt, C.; Bernath, C.; Franke, K.; Pfluger, B.; Willemsen, S.; Sensfuß, F. The role of hydrogen in a greenhouse gas-neutral energy supply system in Germany. *Energy Convers. Manag.* **2022**, *270*, 116188. [CrossRef]
18. Bednarczyk, J.L.; Brzozowska-Rup, K.; Luściński, S. Opportunities and Limitations of Hydrogen Energy in Poland against the Background of the European Union Energy Policy. *Energies* **2022**, *15*, 5503. [CrossRef]
19. Liu, W.; Wen, F.; Xue, Y. Power-to-gas technology in energy systems: Current status and prospects of potential operation strategies. *J. Mod. Power Syst. Clean Energy* **2017**, *5*, 439–450. [CrossRef]
20. Chatterjee, P.; Ambati MS, K.; Chakraborty, A.K.; Chakrabortty, S.; Biring, S.; Ramakrishna, S.; Wong TK, S.; Kumar, A.; Lawaniya, R.; Dalapati, G.K. Photovoltaic/photo-electrocatalysis integration for green hydrogen: A review. *Energy Convers. Manag.* **2022**, *261*, 115648. [CrossRef]
21. Beck, F.J. Rational Integration of Photovoltaics for Solar Hydrogen Generation. *ACS Appl. Energy Mater.* **2019**, *2*, 6395–6403. [CrossRef]
22. Maroufmashat, A.; Fowler, M. Transition of Future Energy System Infrastructure; through Power-to-Gas Pathways. *Energies* **2017**, *10*, 1089. [CrossRef]
23. Temiz, M.; Dincer, I. Design and analysis of a floating photovoltaic based energy system with underground energy storage options for remote communities. *J. Energy Storage* **2022**, *55*, 105733. [CrossRef]
24. Temiz, M.; Javani, N. Design and analysis of a combined floating photovoltaic system for electricity and hydrogen production. *Int. J. Hydrogen Energy* **2020**, *45*, 3457–3469. [CrossRef]
25. Yates, J.; Daiyan, R.; Patterson, R.; Egan, R.; Amal, R.; Ho-Baille, A.; Chang, N. Techno-economic Analysis of Hydrogen Electrolysis from Off-Grid Stand-Alone Photovoltaics Incorporating Uncertainty Analysis. *Cell Rep. Phys. Sci.* **2020**, *1*, 100209. [CrossRef]
26. Nordin, N.D.; Rahman, H.A. Comparison of optimum design, sizing, and economic analysis of standalone photovoltaic/battery without and with hydrogen production systems. *Renew. Energy* **2019**, *141*, 107–123. [CrossRef]
27. Hassan, Q.; Abdulrahman, I.S.; Salman, H.M.; Olapade, O.T.; Jaszczur, M. Techno-Economic Assessment of Green Hydrogen Production by an Off-Grid Photovoltaic Energy System. *Energies* **2023**, *16*, 744. [CrossRef]
28. Sulistyo, H.D.; Setiawan, E.A.; Purwanto, W.W.; Kaharudin, D. Power to Gas-Hydrogen Industry Development Based on Floating PV in Indonesia. In Proceedings of the 2023 13th International Conference on Power, Energy and Electrical Engineering (CPEEE 2023), Tokyo, Japan, 25–27 February 2023. [CrossRef]
29. Li, J.; Gao, C.; Lu, X.; Hoseyni, A. A combined energy system consisting of fuel cell, water electrolyzer and solar technologies to produce hydrogen fuel and electricity. *Energy Sources Part A Recovery Util. Environ. Eff.* **2022**, *44*, 1173–1188. [CrossRef]
30. Song, J.; Choi, Y. Analysis of the potential for use of floating photovoltaic systems on mine pit lakes: Case study at the Ssangyong open-pit limestone mine in Korea. *Energies* **2016**, *9*, 102. [CrossRef]
31. Pouran, H.M.; Padilha Campos Lopes, M.; Nogueira, T.; Alves Castelo Branco, D.; Sheng, Y. Environmental and technical impacts of floating photovoltaic plants as an emerging clean energy technology. *iScience* **2022**, *25*, 105253. [CrossRef] [PubMed]
32. Stachowski, P.; Kraczkowska, K.; Liberacki, D.; Oliskiewicz-Krzywicka, A. Water reservoirs as an element of shaping water resources of post-mining areas. *J. Ecol. Eng.* **2018**, *19*, 217–225. [CrossRef]
33. Kasztelewicz, Z. Approaches to Post-Mining Land Reclamation in Polish Open-Cast Lignite Mining. *Civ. Environ. Eng. Rep.* **2014**, *12*, 55–67. [CrossRef]
34. Kantor-Pietraga, I.; Krzysztofik, R.; Solarski, M. Planning Recreation around Water Bodies in Two Hard Coal Post-Mining Areas in Southern Poland. *Sustainability* **2023**, *15*, 10607. [CrossRef]
35. Indartono, Y.S.; Nur, A.M.; Divanto, A.; Adiyani, A. Design and Testing of Thermosiphon Passive Cooling System to Increase Efficiency of Floating Photovoltaic Array. *Evergreen* **2023**, *10*, 480–488. [CrossRef]

36. Liu, L.; Wang, Q.; Lin, H.; Li, H.; Sun, Q. Power Generation Efficiency and Prospects of Floating Photovoltaic Systems. *Energy Procedia* **2017**, *105*, 1136–1142. [CrossRef]
37. Pavloudakis, F.; Galetakis, M.; Roumpos, C. A spatial decision support system for the optimal environmental reclamation of open-pit coal mines in Greece. *Int. J. Min. Reclam. Environ.* **2009**, *23*, 291–303. [CrossRef]
38. Doležalová, J.; Vojar, J.; Smolová, D.; Solský, M.; Kopecký, O. Technical reclamation and spontaneous succession produce different water habitats: A case study from Czech post-mining sites. *Ecol. Eng.* **2012**, *43*, 5–12. [CrossRef]
39. Güllü, E.; Doğru Mert, B.; Nazligul, H.; Demirdelen, T.; Gurdal, Y. Experimental and theoretical study: Design and implementation of a floating photovoltaic system for hydrogen production. *Int. J. Energy Res.* **2022**, *46*, 5083–5098. [CrossRef]
40. López, M.; Soto, F.; Hernández, Z.A. Assessment of the potential of floating solar photovoltaic panels in bodies of water in mainland Spain. *J. Clean. Prod.* **2022**, *340*, 130752. [CrossRef]
41. Sukarso, A.P.; Kim, K.N. Cooling effect on the floating solar PV: Performance and economic analysis on the case of west Java province in Indonesia. *Energies* **2020**, *13*, 2126. [CrossRef]
42. Kamuyu WC, L.; Lim, J.R.; Won, C.S.; Ahn, H.K. Prediction model of photovoltaic module temperature for power performance of floating PVs. *Energies* **2018**, *11*, 447. [CrossRef]
43. Cao, X.; Wang, J.; Zhao, P.; Xia, H.; Li, Y.; Sun, L.; He, W. Hydrogen Production System Using Alkaline Water Electrolysis Adapting to Fast Fluctuating Photovoltaic Power. *Energies* **2023**, *16*, 3308. [CrossRef]
44. Gül, M.; Akyüz, E. Hydrogen generation from a small-scale solar photovoltaic thermal (PV/T) electrolyzer system: Numerical model and experimental verification. *Energies* **2020**, *13*, 2997. [CrossRef]
45. Shekardasht, S.Z.; Canli, E.; Ates, A. Solar energy supported hydrogen production: A theoretical case study. *Selcuk. Univ. J. Eng. Sci. Technol.* **2017**, *5*, 536–554. [CrossRef]
46. Huang, Y.S.; Liu, S.J. Chinese green hydrogen production potential development: A provincial case study. *IEEE Access* **2020**, *8*, 171968–171976. [CrossRef]
47. First Solaris buses delivered to Bolzano in Italy. *Fuel Cells Bull.* **2021**, 2021. [CrossRef]
48. Key Austrian operator orders 40 Solaris buses. *Fuel Cells Bull.* **2021**, 2021. [CrossRef]
49. Solaris buses order for Frankfurt, Austria trial. *Fuel Cells Bull.* **2021**, 2021. [CrossRef]
50. Solaris hydrogen buses order for Netherlands, Ballard modules order. *Fuel Cells Bull.* **2020**, *2020*, 2–3. [CrossRef]
51. Solaris fuel cell buses for Poland, Germany. *Fuel Cells Bull.* **2021**, 2021. [CrossRef]
52. Solaris hydrogen buses ordered for Sweden. *Fuel Cells Bull.* **2020**, 2020. [CrossRef]
53. Solaris hydrogen buses for NL, Polish demo. *Fuel Cells Bull.* **2021**, 2021. [CrossRef]
54. Kozłowska, O.; Sołomacha, M.; Walentek, I. New data on the resources of sand and grave aggregates for road investment and construction in Poland. *Miner. Resour. Manag.* **2016**, *32*, 103–118. [CrossRef]
55. Martínez-Lage, I.; Vázquez-Burgo, P.; Velay-Lizancos, M. Sustainability evaluation of concretes with mixed recycled aggregate based on holistic approach: Technical, economic and environmental analysis. *Waste Manag.* **2020**, *104*, 9–19. [CrossRef]
56. Szczepiński, J.; Flszer, J.; Stachowicz, Z.; Szczepanik, P. Reclamation of Polish lignite open pits by flooding. *Biul. Panstw. Inst. Geol.* **2010**, *441*, 167–174.
57. Pye, S.; Broad, O.; Bataille, C.; Brockway, P.; Daly, H.E.; Freeman, R.; Gambhir, A.; Geden, O.; Rogan, F.; Sanghvi, S.; et al. Modelling net-zero emissions energy systems requires a change in approach. *Clim. Policy* **2021**, *21*, 222–231. [CrossRef]
58. Panos, E.; Glynn, J.; Kypreos, S.; Lehtilä, A.; Yue, X.; Ó Gallachóir, B.; Daniels, D.; Dai, H. Deep decarbonisation pathways of the energy system in times of unprecedented uncertainty in the energy sector. *Energy Policy* **2023**, *180*, 113642. [CrossRef]
59. Brdulak, A.; Chaberek, G.; Jagodzinski, J. Development forecasts for the zero-emission bus fleet in servicing public transport in chosen EU member countries. *Energies* **2020**, *13*, 4239. [CrossRef]
60. Pardhi, S.; Chakraborty, S.; Tran, D.D.; el Baghdadi, M.; Wilkins, S.; Hegazy, O. A Review of Fuel Cell Powertrains for Long-Haul Heavy-Duty Vehicles: Technology, Hydrogen, Energy and Thermal Management Solutions. *Energies* **2022**, *15*, 9557. [CrossRef]

energies

Article

Battery Electric Roof Bolter versus Diesel Roof Bolter—Results of Field Trials at a Polish Copper Mine

Artur Kozłowski [1] and Łukasz Bołoz [2,*]

[1] Łukasiewicz Research Network–Institute of Innovative Technologies EMAG, Leopolda 31, 40-189 Katowice, Poland; artur.kozlowski@emag.lukasiewicz.gov.pl
[2] Department of Machinery Engineering and Transport, Faculty of Mechanical Engineering and Robotics, AGH University of Krakow, A. Mickiewicza Av. 30, 30-059 Krakow, Poland
* Correspondence: boloz@agh.edu.pl; Tel.: +48-12-617-30-71

Abstract: Battery-powered electric machines have been replacing classic combustion vehicles for many years in the automotive and heavy industry. This change has a positive impact on the environment and, in the case of working machines, also on the safety and comfort of operators. In underground mining plants, due to limited working space and difficult environmental conditions, the use of battery-powered electric vehicles (BEVs) in place of combustion machines with diesel engines brings even greater benefits in terms of the operator's work conditions. This article presents the results of comprehensive tests of two roof bolters in a BEV and a vehicle with a combustion engine. The tests were performed in underground conditions, during normal operation of the machines. They covered many aspects of machines' operation, such as availability; traction properties; battery use; cooling system; efficiency; costs; safety; and ergonomics in terms of gas emissions, noise, vibrations, and generally understood work comfort. The research results showed a significant advantage of the battery-powered machine over the one with a combustion engine. The tests in question are unique due to their scope and the fact that they were carried out in underground conditions, during normal operation, both for the internal combustion machine and its battery-powered equivalent.

Keywords: underground machine; roof bolter; BEV; safety; ergonomics; comfort; field test; exposure to noise; exposure to vibrations; exposure to dust

1. Introduction

In metal ore mines, including KGHM Polska Miedź S.A. (KGHM Polish Copper Plc), exploitation takes place in increasingly difficult conditions, which result primarily from the increasing depth of the seams and their small thickness. The rock temperature and natural hazards increase with the depth of exploitation. Additionally, the smaller thickness of the deposits limits the space, which makes the issue of generated by internal combustion engine machines temperature, noise and gases crucial. To meet these requirements, Mine Master Sp. z o. o. (Ltd., Limited company, Wilków, Poland), in cooperation with AGH University of Krakow and Łukasiewicz Research Network—EMAG Institute of Innovative Technologies, developed, manufactured, and commissioned at KGHM Polska Miedź S.A. (joint-stock company, Lubin, Poland), a self-propelled roof bolter with an electric drive and battery power Roof Master 1.8 KE (Figure 1). Compared to the currently used roof bolters equipped with a combustion engine, the introduction of battery power in underground conditions is a significant step toward improving the atmosphere in the mine. This type of power supply expands the area of application of the vehicles and significantly improves occupational health and safety conditions by eliminating exhaust gases, as well as reducing noise, vibrations, and thermal load in the longwall. It should be emphasised that this machine is intended for low excavations, which is a unique feature compared to competitors' solutions, especially battery ones.

Figure 1. Battery-powered self-propelled Roof Master 1.8 KE roof bolter in the KGHM mine.

Apart from self-propelled roof bolters, the main machines in the room-and-pillar mining system include self-propelled drilling rigs, LHD loaders, and haul trucks. The user expects greater safety and comfort of work from a battery-powered machine than from the one equipped with an internal combustion engine. However, the key assumption is to achieve at least the same technical parameters and operational properties as in the case of traditional machines. This article presents the results of operational tests and experiences with the operation of the battery-powered electric drill rig (Roof Master 1.8 KE). The tests showed its superiority over the combustion version (Roof Master 1.8 AWK) in terms of broadly understood economy, ergonomics, safety, and comfort of work (Figure 2). It is worth emphasising that both machines were equipped with an identical working system for drilling and bolting. This article presents the results of comprehensive tests covering many aspects of these machines' operation, such as work availability, traction properties, functionality, efficiency, maintenance costs, ergonomics, air conditioning system, and the protection devices installed. In addition to the abovementioned tests, the RM 1.8 KE bolting rig has successfully passed all tests and examinations required by national directives and regulations, including those in the field of stability and electromagnetic compatibility.

Figure 2. Self-propelled roof bolters Roof Master 1.8 KE and Roof Master 1.8 AWK produced by Mine Master (source of "vs": Freepik.com).

2. State of the Art

In order to reduce operating costs and improve the conditions of the crew's work, it is advisable to replace diesel engines with main-supplied and battery-powered electric motors. In general, there is a tendency to move away from fossil fuels as an energy source [1]. This trend has been observed in the industry for many years, especially in the case of self-propelled mining machines, such as drilling and bolting rigs, LHD loaders, and haul trucks.

In underground mining, battery-powered machines began to be designed and used relatively late due to very difficult working conditions, high technical requirements, and formal requirements resulting from regulations. Currently, many manufacturers offer solutions for battery-powered machines that are designed to work in various conditions. These are companies such as Aramine, Artisan Vehicles (since 2019 has belonged to Sandvik), Epiroc (until 2018 Atlas Copco), Komatsu (until 2017 Joy Global), MacLean Engineering, Normet, Phillips Machine, Sandvik, Komatsu, and CAT [2]. In the process of designing machines and work equipment, modern methods are used so as to meet the requirements of difficult working conditions. Great importance is attached to numerical tests of cabins in order to obtain the highest possible level of operator safety [3,4]. Similarly, in the field of quality control and production management, modern methods and tools are applied in order to improve the quality of machines and devices [5–9]. However, only tests in real conditions allow the machine to be fully tested and its capabilities and operational properties to be verified.

There are only a few studies and articles concerning the scope of research described in this article. A very interesting and valuable study containing research and analysis results has been presented in one of the articles [10]. The authors have made a comprehensive assessment of battery-powered rigs. In the article, they have presented the change in costs and technical parameters observed in recent years due to the use of batteries. Although the article is from the automotive industry, it concerns the common thread of all BEVs, i.e., batteries. In another article, progress in the field of vibrations of battery-powered vehicles and machines has also been presented [11].

In her study, Grycan addresses the issue of the safe operation of battery machines in underground mining conditions [12]. In this article, she raised the issue of specific working conditions in underground mines and their impact on the safety of operation of battery-powered machines compared to their combustion equivalents.

The authors of another article described a strategy for reducing ventilation and heating costs, which was applied in one of the Swedish underground iron ore mines [13]. The authors demonstrated in their study that the use of battery-powered machines instead of combustion ones allows for a reduction in the total costs of ventilation and heating by 36%.

A very broad study is an article on improving the working conditions in an underground mine in Finland, which was achieved thanks to the use of battery-powered machines instead of internal combustion ones [14]. The research concerned several types of machines (a haul truck, LHD loader, and drilling rig). The authors carried out measurements of gas emissions as well as dust and air temperature increases. Additionally, they conducted surveys among both machine operators and employees working around these machines so as to obtain their opinions. The results of the survey indicate a significant improvement in working conditions owing to the use of BEVs.

Similarly, Professor Szlązak and co-authors demonstrated in their article that the use of battery-powered suspended railways has a beneficial effect on working conditions in mining excavations [15]. The authors analysed air quality (elimination of exhaust gases) and climatic conditions (reduction in air temperature and humidity). Their article presents the results of basic research on internal combustion machines and the expected effects of introducing battery-powered trains. The abovementioned articles mainly concern the impact of the use of BEVs on air quality and temperature in underground mine workings. All studies indicate improved conditions.

As regards the exposure of machine operators to vibrations and noise generated by combustion and electric machines, individual studies can be found, but they usually concern machines used in open-pit mining. An example is the research described in the article by Tekin regarding operators of internal combustion machines in Turkey in the field of WBV (whole-body vibration) [16]. Similarly, the article by Erdem et al. describes WBV observed in haul truck operators [17]. The topic of noise generated by various devices, including underground mining machines, both diesel and electric, has been raised in an article on mining processing machines [18]. Due to the wide scope of BEV applications in the automotive industry and heavy industry, there are numerous research results and studies on the benefits of using battery-powered machines in comparison with their classic combustion equivalents [19–25].

This article presents the results of comprehensive comparative tests of an internal combustion machine and its electric, battery-powered counterpart. The tests were carried out in a very wide scope, but they concerned one type of self-propelled mining machine, i.e., the roof bolter. It should be noted that research conducted in an underground mine is expensive, complicated, and organisationally demanding; thus, the number of available publications in this field is significantly limited, and the existing ones concern only selected aspects of the machines' work.

3. Materials and Methods

The conducted comprehensive in-service tests consisted of many elements. They were based on data recorded by the machine, mine reports, and the results of the performed research. When the machine with an internal combustion engine is replaced by a battery-powered one, it seems natural and reasonable to compare the two solutions. The battery machine can be described using a number of parameters related to various aspects of its operation, such as range, operating time, efficiency, generated noise, or vibrations. These parameters can be compared to those obtained by combustion machines. Such an approach allows us to compare the machines in an objective way and draw clear conclusions. Therefore, conducting tests under normal operating conditions in an underground mine is an additional advantage. Hence, the results and conclusions from underground tests of the Roof Master 1.8 KE battery-powered roof bolter were compared to the results and conclusions from the tests of the Roof Master 1.8 AWK diesel machine. Both manufactured by Mine Master. On the basis of mine reports, the changes made to the machine were described, and its availability as well as the consumption of utilities and consumables were determined.

Throughout the period of in-service testing, a number of parameter values collected by the machine were recorded thanks to the use of a modern control and diagnostic system. The information recorded by the machine includes, among others, distance; the condition and consumption of the battery; the number of charging cycles; electrical parameters of the motors; and temperatures, including those in the operator's cabin. Based on this information, the average number of bolts per shift, average distance per shift, and other issues related to ergonomics and work safety were calculated.

The tests requiring separate measurements with the use of external sensors, or the ones installed in the machine for the purposes of these tests were as follows:

- Measurements of chemical factors (nitrogen oxide and nitrogen dioxide) in the operator's cabin;
- Measurements of dust (unclassified dust and silica) in the operator's cabin;
- Measurements of vibrations affecting the operator sitting on a cushioned seat while driving and installing a bolt;
- Measurements of noise affecting the operator in the cabin and the staff outside the machine;
- Measurements of the effectiveness of the machine's braking system (speed, distance, braking delay, and excavation slope).

The above tests were performed in accordance with relevant standards and recommendations with the use of specialised sensors and measurement systems. Measurements and calculations of the workers' exposure to noise load were made in accordance with the PN-EN ISO 9612:2011 standard [26] (Acoustics—Determination of Occupational Noise Exposure–Engineering Method). The methodology as well as sample measurements and calculations were presented in another article by one of the authors [18]. Measurements and calculations regarding the braking system were carried out in accordance with the PN-EN ISO 3450:2011 standard [27] (Earth-Moving Machinery—Wheeled or High-Speed Rubber-Tracked Machines—Performance Requirements and Test Procedures for Brake Systems). Measurements and calculations of vibrations to which employees are exposed were made according to the PN-EN 14253:2011 standard [28] (Vibration and Shock with Respect to Human beings).

Throughout the period including test runs and in-service tests, all machine parameters were recorded. All of the tests described in this article were performed in one of the KGHM mines where the machine was tested. Examples of photos of the operator's cabin and roof drilling taken during the tests are shown in Figure 3. In the article, charts for selected rides or selected days are presented. However, the calculated values and the presented conclusions apply to the entire testing period.

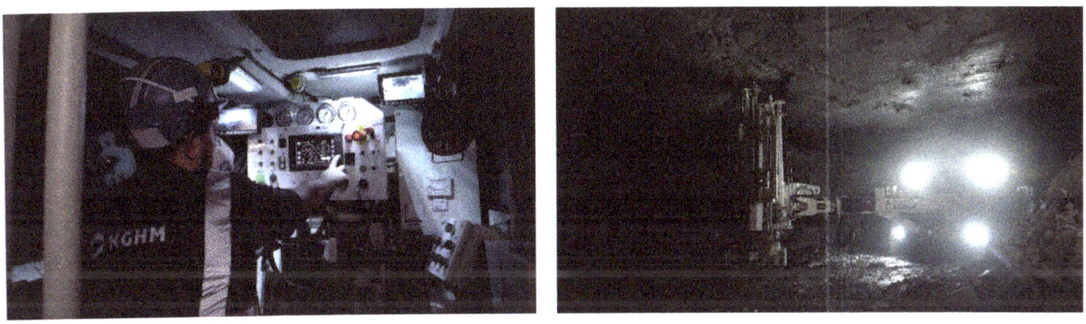

Figure 3. Photos of the Roof Master 1.8 KE roof bolter taken during the tests.

In addition to the measured values that allow for a quantitative description of the machine's parameters, the opinions of experienced machine operators were also taken into consideration. These opinions were collected in the form of reports that they created when using the machine. A crucial issue was also to use a simple but very effective method of collecting information, i.e., a direct interview. The direct effect of the direct interview was a number of proposals for changes and improvements, which were applied and are also mentioned in this article.

4. Underground Field Tests

The priority assumption for in-service tests of the Roof Master 1.8 KE and Roof Master 1.8 AWK roof bolters was to carry them out at the target workplace, which makes these tests unique. However, this involved a number of difficulties resulting from the need to ensure a normal operation of these machines and other machines working at that time and place.

4.1. Description of the Research Object

This research concerned the Roof Master 1.8 KE roof bolter. It should be clearly emphasised that this is the world's first battery-powered machine with built-in chargers, intended for low excavations. This roof bolter has a height of 1.8 m, and its design and dimensions allow for manoeuvring in perpendicular excavations with a width of 4.4 m. Solutions offered by competitive companies require excavations much larger than 2.0 m. Battery chargers installed in the machine enable direct charging from mine mains without the need to use additional infrastructure. Additionally, the machine has a braking energy recovery system for recharging the battery. It is a typical BEV (battery electric vehicle) machine equipped with a pack of 5 HV (high voltage) sodium–nickel batteries with a capacity of 190 Ah and energy of 123.5 kWh, at a constant nominal voltage of 650 V DC (Direct Current). It is driven by one permanent magnet synchronous motor (PMSM) with a power of 133 kW. The motor drives the torque converter, gearboxes, and axles through the drive shafts on one side, and on the other, the hydraulic unit and the air conditioning compressor. The machine is adapted to work in power grids with rated voltages of 3 × 500 V, 3 × 690 V, and 3 × 1000 V, with a rated frequency of 50 Hz AC (alternating current).

The RM 1.8 KE roof bolter is designed for roof bolting in workings with a height of 3.0 m to 5.9 m. The dimensions of the machine allow it to move in galleries with a minimum height of 2.0 m. The machine can move in workings with a longitudinal inclination of up to 15° and a transverse inclination reaching up to 8°. It is equipped with an AWK-1.8 drilling and bolting turret, which allows for the installation of expansive and adhesive bolts. The kinematic system of the telescopic boom enables bolts to be installed in the roofs and sidewalls. The AWK-1.8 automatic turret has a rotating clip for 9 bolts. Drilling is performed with the use of a hydraulic rotary drill having a torque of 450 Nm. Drilling rods with a length of 2160 mm have been used as standard [25].

4.2. Place and Time of In-Service Tests

In-service tests were carried out at KGHM Polska Miedź S.A., in the Lubin mining division, in the period from 31 December 2020 to 30 April 2022. The aim of the tests was to confirm the technical assumptions and suitability of the battery-powered machine for use in KGHM copper ore mines, especially in relation to combustion machines.

During in-service tests, the machine was used like any other machine in motion. Throughout the testing period, it installed bolts and made additional holes in the following amounts:

- Expansion bolts with a length of 1.8 m: 13,853 pcs;
- Adhesive bolts with a length of 1.8 m: 52 pcs;
- Additional bolts for mounting accessories: 28 pcs;
- Additional holes made: 243.

4.3. Machine Availability

Throughout the testing period, the RM 1.8 KE machine worked for a total of 49 weeks, during which 943 machine shifts (Tn) were recorded. There were periods in which the machine worked and was operational—these were described by the number of machine shifts for the operational machine (PRD), as well as periods when the machine was operational in planned readiness mode (DPL) and unplanned readiness mode (DNP).

During the analysed 49 weeks, the machine achieved the following parameters:

- Number of machine shifts Tn = 943.0;
- Number of machine shifts for operational machine (PRD) = 411.5;
- Number of machine shifts in planned readiness mode (DPL) = 327.3;
- Number of machine shifts in unplanned readiness mode (DNP) = 5.0.

It can therefore be calculated that the machine has achieved the following technical availability (Dt):

$$\mathrm{Dt} = \frac{(\mathrm{PRD} + \mathrm{DPL} + \mathrm{DNP})}{\mathrm{Tn}} \times 100\% = 78.88\% \tag{1}$$

In the analysed period, the machine was subjected to various modifications aimed at ensuring greater safety as well as easier operation to improve its efficiency. The changes were proposed by experienced operators (mine workers) and concerned various elements of the machine. They included the following features:

- Addition of extra charging diodes;

- Addition of warning flashing lights;
- Modification of the control of water and cable reels;
- Software update;
- Replacement of wires;
- Relocation of the expansion tank;
- Improvement in distributors' availability;
- Addition of extra shields (Figure 4a);
- Addition of foldable steps (Figure 4b);
- Change in camera position;
- Installation of an additional camera (Figure 4c).

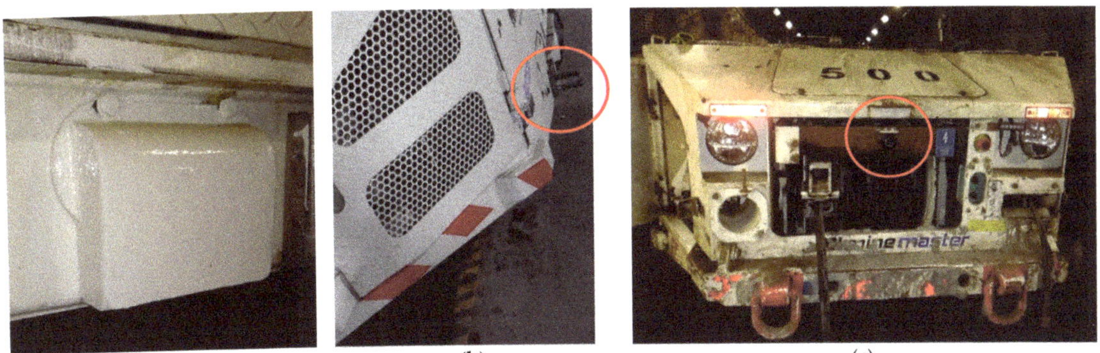

(a) (b) (c)

Figure 4. Selected modifications of the machine: (**a**) addition of a hot air outlet cover; (**b**) addition of foldable steps (in the red circle); (**c**) installation of an additional camera (in the red circle).

The availability of the machine was influenced by the need to introduce the abovementioned changes and the very fact of carrying out measurements, especially with the use of additional equipment. In total, the modifications made during the machine's operation took 55 machine shifts. Additional downtimes resulted from the need to prepare the machine and carry out tests, which took a total of seven machine shifts. Given these additional circumstances, the actual machine availability was calculated to reach 84.43%.

4.4. Traction Properties and Battery Charge Level

During test runs, the machine was driven from the chamber to the place of work and back. The average distance covered per shift ranged from 6 km to 7 km, and the average battery consumption was approximately 6% per 1 km. The RM 1.8 KE machine climbed hills with an inclination of approximately 10° at a speed of 4 km/h, which confirmed the design assumptions. The Michelin tubeless tires (12.00-24 XZM TL 178A5 NXZM) and the standard hydrodynamic drive system (gearbox and torque converter), despite an increase in the machine weight to 24,590 kg, did not result in any deterioration of the traction parameters compared to RM 1.8 AWK.

The drawings below show graphs illustrating the parameters recorded during one of the machine's journeys to the workplace, as well as its return to the chamber. During the journey to the workplace, the vehicle covered 3.5 km, and the inclination varied in the range of +/− 10°. The machine used 16% of the energy stored in the battery. Initially, the battery charge level (state of charge—SoC) reached 79%, and the final value was 63%. On average, 4.7% of battery energy was used by the machine to cover 1 km (Figures 5 and 6).

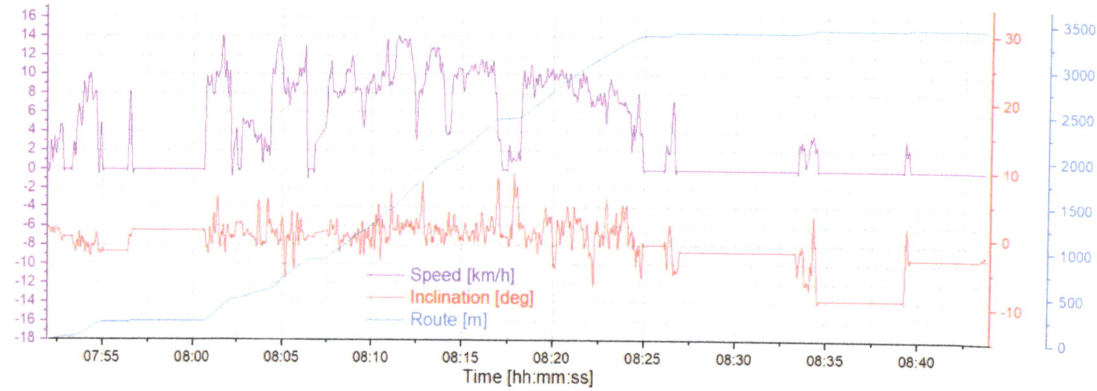

Figure 5. Longitudinal inclination, speed, and travel distance of the roof bolter from the longwall to the chamber.

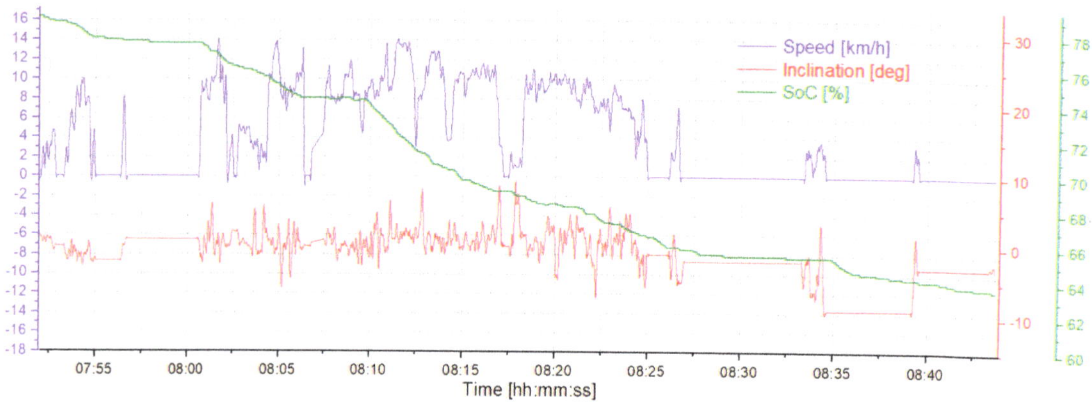

Figure 6. Longitudinal inclination, speed, and the level of charge of the roof bolter travelling from the chamber to the longwall.

The distance covered by the machine from the longwall to the inspection–repair chamber was 2.95 km. The mileage recorded for this route is shown in Figures 7 and 8. While driving, the machine used 21% of the battery energy. Initially, the battery charge reached 73%, and the final value was 52%. Therefore, the machine consumed an average of 7% of battery charge per 1 km.

Given the fact that the average discharge of the HV traction battery during the tests was approximately 6%/1 km and that the battery charge should not fall below 20%, the machine can travel approximately 13 km without recharging the battery in mining conditions. Additionally, the entire process of refuelling the machine, which occurs in the case of machines with an internal combustion engine, has been eliminated.

The tests showed that after completing work related to the installation of a roof bolting support, the machine returned to the repair chamber with the set of batteries charged to approximately 50%, which is significantly higher than the minimum required battery charge level. The traction properties of the battery-powered machine were in no way inferior to those achieved by machines with an internal combustion engine.

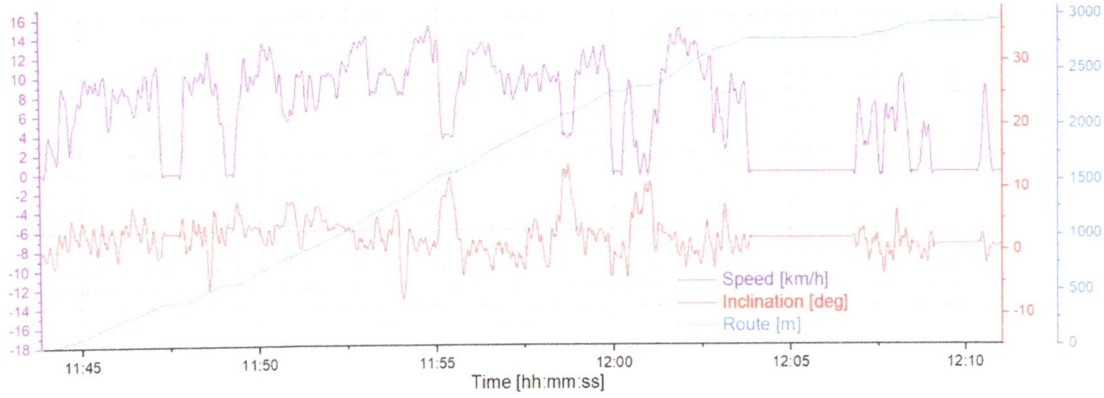

Figure 7. Longitudinal inclination, speed, and travel distance of the roof bolter from the longwall to the chamber.

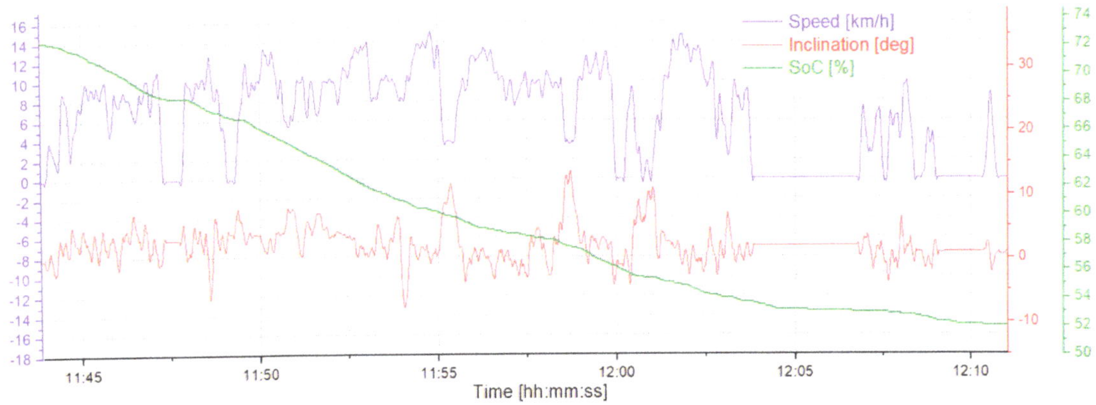

Figure 8. Longitudinal inclination, speed, and the level of charge of the roof bolter travelling from the longwall to the chamber.

4.5. Efficiency of the Drilling and Bolting Process

The purpose of using self-propelled roof bolters is to install a roof bolting support. Therefore, any action that allows for achieving a higher bolting efficiency is desirable. One such action was the use of electrically controlled distributors, which enabled increasing the dynamics of the working system control and adapting it to the user's needs.

During the tests, the machine's efficiency was assessed positively. The average number of 1.8 m long expansion bolts installed per shift reached 36.44, which is more than in the case of roof bolters with a combustion drive of 33.76 bolts/shift.

4.6. Work Safety

One of the key issues related to the safety of mining trucks is the service braking system and the emergency brake (EB). Table 1 summarises the results of tests and measurements of the machine speed and braking effectiveness carried out at the end of the tests of the RM 1.8 KE machine in underground conditions. The tests and calculations were based on the formulas from the PN-EN ISO 3450:2011 standard and Annex A to this standard. The results include the maximum speed [km/h], braking distance s [m], and braking delay a [m/s^2]. Tests and calculations were performed for all three gears and the service brake, and, additionally, for gear 1 and the emergency brake. Figure 9 shows a photo from braking efficiency tests on a slope of 12°. Braking speed measurements, carried out in accordance with applicable regulations, did not reveal any abnormalities in the operation of the suspension system.

Table 1. Results of speed and braking measurements for RM 1.8 KE.

Braking	v [km/h] Speed	s [m] Measured	s [m] According to Standard 3450	s [m] According to Annex A	a [m/s^2] Measured
Gear 1 Service brakes	4	0.4	1.90	0.61	4.23
Gear 2 Service brakes	10	1.5	3.63	2.37	5.58
Gear 3 Service brakes	15	3.3	5.41	4.61	7.69
Gear 1 Emergency brake EB	4	0.6	3.80	1.46	3.73

Figure 9. Photo from braking effectiveness tests on a slope of 12°.

4.7. Ergonomics and Comfort of Work

From the beginning of the design process of the Roof Master 1.8 KE roof bolter, one of the most important assumptions was to meet the requirements and expectations in the field of broadly understood ergonomics. The most important process is the installation of a roof bolting support, which is controlled by the operator from the cabin. Therefore, the ability to observe the working machine from the cabin is crucial. The cabin has good visibility and optimal dimensions enabling comfortable work (Figures 10 and 11). Passage height amounts to 1800 ± 30 mm. Cabin is raised by 270 mm.

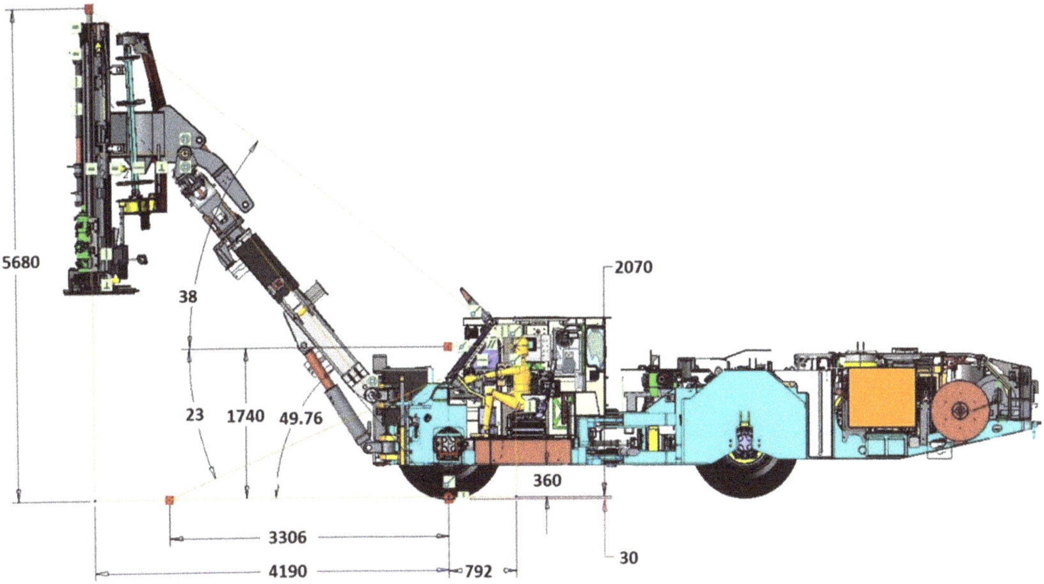

Figure 10. Geometric parameters and field of vision of the cutting head.

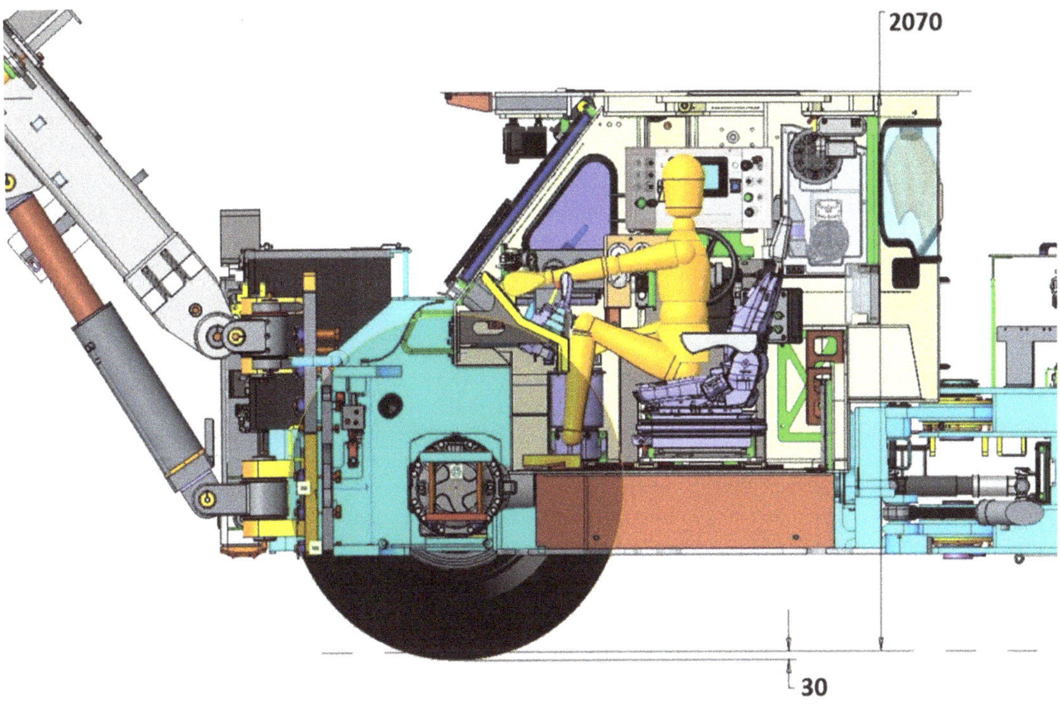

Figure 11. Operator's position in the cabin during work.

Problems with limited visibility reported by the user when driving the machine with the tractor facing forward were solved by applying a larger, 12-inch screen displaying the image from the camera (Figure 12). During work, visibility was improved using a hydraulically raised canopy segment.

Figure 12. View from the camera on a 12″ display (the green lince—center-vertical line).

Users emphas0ised that thanks to the battery power and an electric motor used in place of an internal combustion engine, exhaust gas emissions were completely eliminated, which significantly improved the working conditions for the operator, service staff, and people staying in the area. In the battery-powered machine, the operators felt that heat emission was reduced owing to the lack of a diesel engine and a high-temperature exhaust system, as well as due to the relocation of the hydraulic distributors outside the cabin. Research on the exposure of operators and service technicians of the RM 1.8 KE and RM 1.8 AWK to exhaust emissions, noise, and vibrations was carried out. Examples of dust and gas test results are presented in Table 2, whereas Table 3 presents the results of measurements of nitrogen oxide and dioxide in the cabin.

In addition to measuring the sound level, noise tests require a number of calculations to determine the equivalent sound level (Tables 4 and 5). The measured and calculated equivalent sound level A $L_{Aeq,Te}$ for RM 1.8 KE (battery) was 76 dB, while for the RM 1.8 AWK (combustion engine), it reached as much as 86 dB. This means that the battery-powered machine is ten times quieter (log-scale). Such a significant noise reduction required modifying the machine by adding warning lighting in order to make it visible to other workers. The vibrations induced by the drive system during the machine's travel were 0.372 m/s^2 for the RM 1.8 KE machine and 0.393 m/s^2 for RM 1.8 AWK, which reduces the operator's fatigue and has a positive impact on his well-being (Table 6).

Table 2. Results of measurements of dust and silica in the cabin.

Determined Chemical Agent	NDS [mg/m^3]	Measurement Duration [min]	C [mg/m^3]	Cw ± Uc [mg/m^3]
Dust not classified in terms of toxicity—inhalable fraction	10	450	0.13	0.12 ± 0.03
Crystalline silica—respirable fraction	0.1	450	<0.01	<0.01

Table 3. Results of measurements of nitrogen oxide and dioxide in the cabin.

Determined Chemical Agent	NDS [mg/m^3]	Measurement Duration [min]	GG or GGw [mg/m^3]	DG or DGw	Xg or Xgw ± Uc [mg/m^3]	Multiplicity of NDS Value (GG or GGw)	Multiplicity of NDS Value (DG or DGw)	Multiplicity of NDS Value Xg or Xgw	NDSCh [mg/m^3]	Value Comparable with NDSCh ± U_c [mg/m^3]	Multiplicity of NDSch Value
Nitric oxide	3.5	360	3.18	0.56	1.34 ± 0.39	0.9	0.2	0.4	7	6.47 ± 1.04	0.9
Nitrogen dioxide	0.7	360	<0.21	<0.21	0.21	<0.3	<0.3	<0.3	1.5	<0.21	<0.1

Table 4. Noise in the Roof Master 1.8 KE cabin.

Technological Activity	Time of Exposure [min]	Measured Equivalent Level of Sound A $L_{Aeq,Te}$ [dB]	Measured Maximum Level of Sound A $L_{A\,max}$ [dB]	Measured Peak Level of Sound C $L_{c\,peak}$ [dB]	Contribution of Activities to the Level of Exposure $L_{EX,\,8h}$ [dB]
Travel	60	76.0	91.5	126.0	67.0
Work	240	75.7	81.7	103.7	72.7
Remaining time	180	66.0	72.6	100.8	61.8

Table 5. Determined and permissible values of noise in the Roof Master 1.8 KE cabin.

Determined Values of Quantities Characterising Noise			
Daily exposure to noise $E_{A,Te}$ [Pa^{2}s]	288		
	Level of noise exposure $L_{EX,8h}$	Maximum level of sound A $L_{A\,max}$	Peak sound level C L_{Cpeak}
Values of noise level [dB]	74.0	91.5	126.0
Permissible Values of Quantities Characterising Noise			
NDN values: For daily exposure to noise $E_{A,Te}$ [Pa^{2}s]	3640		
	For noise exposure level $L_{EX,8h}$	For maximum sound level A $L_{A\,max}$	For peak sound level C L_{Cpeak}
Po of noise levels [dB]	85.0	115.0	135.0
Trigger threshold values: For quantities characterising noise [dB]	80.0	-	135.0
Risk exposure rating/NDN exceedance multiplicity	0.08 NDN		

Table 6. Vibrations in the Roof Master 1.8 KE cabin during travel.

Direction of vibrations	Equivalent effective frequency-weighted vibration acceleration for the exposure time		Average time of exposure [min]	Effective vibration acceleration $[m/s^2]$
	Measured $[m/s^2]$	after taking into account the kx, ky, and kz slope $[m/s^2]$		
Operator's seat				
X	0.266	0.372	480	0.372
Y	0.129	0.181		
Z	0.344	0.344		
Determined value of daily exposure				
Daily exposure expressed as energy equivalent for 8 h of the effective frequency-weighted vibration acceleration dominant among the vibration accelerations determined for the three direction components awx, awy, and awz $[m/s^2]$				0.372
Permissible values of mechanical vibrations with a general effect on the human body				
NDN value:	Daily exposure value $[m/s^2]$			0.80
Trigger threshold value: exposure value $[m/s^2]$				0.50
Determined risk exposure rating (NDN exceedance multiplicity):				0.47

The use of a closed cabin with efficient air conditioning translates into the operator's improved efficiency and concentration, and, in consequence, into his work results. Additionally, the closed cabin separates the cooled interior from all possible heat radiators, such as the front drive axle or hydraulic distributors.

In the opinion of the workers operating the machine during travel and during work, the machine is convenient and comfortable; it has intuitively and ergonomically arranged controls, a comfortable seat, and an efficient air conditioning system, which reduces the operator's fatigue and has a positive impact on his well-being.

4.8. Maintenance Activities

As a result of replacing the combustion drive with an electric one, the maintenance activities required for the machine have fundamentally changed. Additional procedures related to the operation, control, and inspection of the electrical system have been introduced. Daily and periodic inspections do not require special qualifications or authorisation of service personnel. It is worth noting that the batteries and the main components of the electrical system are maintenance-free, except for control and measurement works resulting from the regulations applicable in this area of the mine. The applied HV traction batteries require a periodic equalising charge. The request to perform such charging is signalled on the driving panel. In the event of a failure, it is necessary to use the diagnostic computer provided by the service centre. In general, however, maintenance activities have been significantly simplified in comparison with the internal combustion machine, which has a positive impact on availability and operating costs.

4.9. Estimated Consumption of Energy Utilities

All information regarding utilities consumed, i.e., electricity for the Roof Master 1.8 KE battery-powered machine as well as fuel and electricity for the Roof Master 1.8 AWK combustion machine, was collected during in-service tests. For both machines, the number of changes, the distance covered, and the number of bolts installed were recorded. Due to the availability of information, the machines were compared on the basis of data from two months of their continuous operation. The collected and calculated data are summarised in Table 7.

Table 7. Estimated consumption of energy utilities.

No.	Item	Unit	RM 1.8 AWK	RM 1.8 KE
1	Total number of bolts installed	[pcs]	3646	2551
2	Number of bolts installed per 1 shift	[pcs/1 shift]	33.76	36.44
3	Total energy consumed	[kWh]	5169	7723
4	Total diesel oil consumed	[dm^3]	2364.67	0.00
5	Average fuel consumption	[dm^3/1 shift]	21.90	0.00
6	Total distance covered	[km]	738	376
7	Average distance covered during 1 shift	[km/1 shift]	6.83	5.36
8	Number of working shifts for data sample	[shift]	108	70

Based on the average cost of diesel oil and electricity, values of selected indicators per bolt were calculated. At the time of testing, the average cost of diesel oil amounted to PLN 4.73 /1 dm^3 (approximately 1.1 Euro/1 dm^3), while the average cost of electricity was PLN 0.62/1 kWh (approximately 0.15 Euro/1 dm^3). The obtained results are summarised in Table 8.

Table 8. Selected indicators describing the cost of bolt installation.

No.	Item	Unit	RM 1.8 AWK	RM 1.8 KE
1	Cost of electrical energy consumed	[PLN]	3256	4865
2	Cost of diesel oil consumed	[PLN]	11,185	0.00
3	Average consumption of electrical energy	kWh1/1 bolt	1.42	3.03
4	Average consumption of diesel oil	[dm^3/1 bolt]	0.65	0.00
5	Average cost of electrical energy consumption	[PLN/1 bolt]	0.89	1.91
8	**Average cost of energy**	**[PLN/1 bolt]**	**3.96**	**1.91**

The average cost of energy per bolt amounted to 1.91 PLN for the battery bolter and 3.96 PLN for the combustion bolter. Thus, the cost of energy per bolt was more than twice less for the battery-powered machine. It should be noted that the consumption of electrical energy by the combustion machine is related to the use of the electric motor of the hydraulic unit, which is responsible for the working system.

4.10. Functionality in the Opinion of Operators—Direct Interview

A direct interview with battery machine operators who had previously worked with internal combustion machines resulted in a number of conclusions. During test runs, the users emphasised good operating parameters and higher bolting efficiency than equivalent machines with an internal combustion drive. The hydraulic system of the Roof Master 1.8 KE machine, equipped with a battery drive, is more efficient than standard Roof Master 1.8 AWK rigs, which is clearly noticeable in terms of the dynamics and smoothness of the working system movements.

As regards operation and maintenance, the users emphasised that the operating system and drive transmission applied in the battery-powered machine correspond to those used in earlier internal combustion machines known to operators and mechanics, which allowed for an efficient and trouble-free change-over to the battery-powered machine, without additional training. Moreover, the use of electrohydraulic distributors, i.e., electric control, eliminated all hydraulic connections between the operator panel and the distributors, thus enabling better maintenance service.

The undoubted advantage of the battery-powered machine is the possibility of charging its batteries without an additional charger or modification of the power grid used at KGHM. The machine can charge the battery directly from standard sockets in the mine's power grid. The current drawn from the network is set and limited automatically; the machine is designed for multi-voltage operation.

A useful and appreciated functionality is the possibility of installing a roof bolting support solely from battery power, without the need to unwind the cable and connect the machine to the mine's mains. Due to the fact that the consumption of energy from the battery reached approximately 3% of the charge level per bolt, such activities can be carried out to a limited extent. It is worth noting that the quoted decrease in the battery charge level applies to the entire process, i.e., drilling a hole, taking a bolt from the magazine, and tightening the bolt with appropriate torque.

5. Summary and Comparison of Factors Affecting the Employee

Measurements of factors affecting the employee during work were performed for both the battery-powered machine (RM 1.8 KE) and the one with an internal combustion engine (RM 1.8 AWK). The values of the processed results for both machines are summarised in the tables below. Table 9 contains the test results for unclassified dust and crystalline silica as well as for nitrogen oxide and dioxide. Table 10 lists the results of tests of vibrations affecting the operator. Table 11 shows the noise test results.

In the vast majority of analysed cases, the battery-powered electric machine compares favourably to the machine with an internal combustion engine—it is usually significantly better.

Table 9. Comparison of employee exposure to dust and gases.

No.	Factors	RM 1.8 AWK	RM 1.8 KE
1	Dust not classified in terms of toxicity, inhalable fraction [mg/m^3]	0.86 ± 0.22	0.12 ± 0.03
2	Crystalline silica—quartz, cristobalite—respirable fraction [mg/m^3]	<0.01	<0.01
	Nitric oxide		
3	GG or GGw [mg/m^3]	3.18	1.98
4	DG or DGw [mg/m^3]	0.56	0.41
5	Xg or Xw ± Uc [mg/m^3]	1.34 ± 0.39	0.90 ± 0.26
6	Value comparable with NDSCh ± Uc [mg/m^3]	7.18 ± 1.15	4.39 ± 0.70
	Nitrogen dioxide		
7	GG or GGw [mg/m^3]	<0.21	<0.21
8	DG or DGw [mg/m^3]	<0.21	<0.21
9	Xg or Xw ± Uc [mg/m^3]	<0.21	<0.21
10	Value comparable with NDSCh ± Uc [mg/m^3]	<0.21	<0.21

Table 10. Comparison of worker exposure to vibrations.

	Vibrations	RM 1.8 AWK	RM 1.8 KE
1	Average time of employee exposure to vibrations in the process of bolting roofs and sidewalls during one working shift.	280	280
2	Calculated value of the frequency-corrected vibration acceleration dominant among the vibration accelerations determined for three direction components X, Y, and Z in the process of bolting roofs and sidewalls [m/s^2]	0.18	0.13

Table 10. *Cont.*

	Vibrations	RM 1.8 AWK	RM 1.8 KE
3	Average time of employee exposure to vibrations during a work shift while travelling to the place of work [min]	60	60
4	Calculated value of the frequency-corrected vibration acceleration dominant among the vibration accelerations determined for three direction components X, Y, and Z during the journey to the place of work [m/s^2]	0.73	0.78
5	Daily exposure expressed as the energy equivalent for 8 h of effective frequency-corrected vibration acceleration dominant among the vibration accelerations determined for three direction components, taking into account the appropriate coefficients [m/s^2]	0.31 ± 0.06	0.31 ± 0.07
6	Short-term exposure lasting 30 min or less, expressed as the energy equivalent of effective frequency-corrected vibration acceleration dominant among the vibration accelerations determined for three direction components, taking into account the appropriate coefficients [m/s^2]	0.45 ± 0.13	0.50 ± 0.14

Table 11. Comparison of employee exposure to noise.

	Noise	RM 1.8 AWK	RM 1.8 KE
1	Operation of a self-propelled roof bolter: driving from the parking place to the site of work, returning [dB]	Maximum sound level: 95.2	Maximum sound level: 92.0
2		Peak sound level: 120.3	Peak sound level: 124.1
3		Contribution of activities to daily exposure level: 80.0	Contribution of activities to daily exposure level: 70.9
4	Operation of a self-propelled roof bolter: when bolting roofs and sidewalls [dB]	Maximum sound level: 93.5	Maximum sound level: 90.8
5		Peak sound level: 118.4	Peak sound level: 108.5
6		Contribution of activities in daily exposure level: 81.1	Contribution of activities in daily exposure level: 79.7

6. Conclusions

The abovementioned tests of the Roof Master 1.8 KE machine provided an adequate amount of data and information on the basis of which the suitability of the machine for work in underground mining plants can be clearly assessed. The analysis results indicate that the use of battery power has a positive effect in terms of all the considered features. The design assumptions regarding the selection of the battery-powered machine in place of the machine with a combustion engine, during its normal work in real conditions of an underground mine, were confirmed.

Previous experiences with the use of battery drives in mines where such drives were implemented point to the following benefits:

- Elimination of exhaust gas emissions (CO_2 and NOx), which in some cases enabled reducing ventilation costs;
- Lower operating costs compared to machines powered by diesel engines;
- Reduced operation and maintenance costs;
- Two-fold reduction in the average cost of energy equivalent for installing one bolt;
- Reduced heat emission;
- Improved work comfort by reducing vibrations and noise;
- Reduced maintenance costs;
- Elimination of costs related to the transport and distribution of liquid fuels.

Moreover, it is worth mentioning a few advantages and comments highlighted by the users. The machine can charge the battery directly from standard sockets in the mine's power grid without the need to use an additional charger. A useful functionality of the machine revealed during the tests was the possibility of installing a roof bolting support only with the use of batteries. The station for driving and working with the machine is identical to that for machines equipped with an internal combustion drive, which makes it easier for operators and mechanics to smoothly switch from internal combustion to battery-powered machines. The design of the working system does not differ from those previously used, which facilitates operation and maintenance activities. The batteries and the main components of the electrical system are maintenance-free, except for control and measurement works resulting from the regulations applicable in this area in the mining plant.

The use of a battery power supply in the machine enables eliminating exhaust emissions and reducing heat. Additionally, a reduced level of noise and vibration translates into higher, noticeable comfort of the operator's work. Thanks to a closed, isolated, and efficiently air-conditioned cabin, better working conditions and comfort were achieved.

The results obtained during the tests indicate that it is advisable to put such machines into operation, which, combined with clean energy, will consequently lead to obtaining a mine product with the lowest possible carbon footprint and a higher level of comfort and safety.

Author Contributions: A.K. analysed the literature and state of knowledge. Ł.B. planned the methodology of the studies. A.K. and Ł.B. performed the tests. A.K. elaborated on the research results. Ł.B. analysed the research results. A.K. and Ł.B. analysed and interpreted the results. Ł.B. elaborated on the results. All authors have read and agreed to the published version of the manuscript.

Funding: This research was conducted with the support of the statutory work project number POIR.01.01.01-00-D011/16 "Nowa generacja modułowych maszyn, wiercącej i kotwiącej, z napędami bateryjnymi, przeznaczonych do pracy w podziemnych kopalniach rud miedzi i surowców mineralnych" ("New generation of battery-powered modular drilling and bolting machines intended for work in underground mines of copper ores and mineral resources").

Institutional Review Board Statement: Not applicable.

Informed Consent Statement: Not applicable.

Data Availability Statement: Data are contained within the article.

Acknowledgments: Many thanks to Mine Master Spółka z o.o. for providing materials and cooperation resulting in the creation of modern battery-powered machines. Thanks to KGHM Polska Miedź S.A. for enabling and supporting the underground tests.

Conflicts of Interest: The authors declare no conflicts of interest.

References

1. Midor, K. Analysis of the Readiness of Silesian City Inhabitants for Decarbonisation. *Acta Montan. Slovaca* **2022**, *27*, 360–367. [CrossRef]
2. Kozłowski, A.; Bołoz, Ł. Design and Research on Power Systems and Algorithms for Controlling Electric Underground Mining Machines Powered by Batteries. *Energies* **2021**, *14*, 4060. [CrossRef]
3. Działak, P.; Karliński, J.; Rusiński, E. Method of Operator Safety Assessment for Underground Mobile Mining Equipment. *E3S Web Conf.* **2018**, *29*, 00009. [CrossRef]
4. Karlinski, J.; Ptak, M.; Dzialak, P. Improving Occupant Safety in Heavy Goods Vehicles—Universal Energy Absorbing System. In Proceedings of the 2018 XI International Science-Technical Conference Automotive Safety, Žastá, Slovakia, 18–20 April 2018.
5. Midor, K.; Wilkowski, G. Recertification of a Quality Management System Based on ISO 9001—Is It a Must for a Modern Manufacturing Company? *Prod. Eng. Arch.* **2021**, *27*, 217–222. [CrossRef]
6. Midor, K.; Kučera, M. Improving the Service with the Servqual Method. *Manag. Syst. Prod. Eng.* **2018**, *26*, 60–65. [CrossRef]
7. Zasadzień, M.; Midor, K. Statistical Process Control as a Failure Removal Improvement Tool. *Acta Technol. Agric.* **2018**, *21*, 124–129. [CrossRef]
8. Zasadzień, M. Using the Pareto Diagram and Fmea (Failure Mode and Effects Analysis) to Identify Key Defects in a Product. *Manag. Syst. Prod. Eng.* **2014**, *16*, 153–156. [CrossRef]
9. Zasadzień, M.; Midor, K. Innovative Application of Quality Management Tools in a Hard Coal Mine. In Proceedings of the 15TH International Multidisciplinary Scientific Geoconference SGEM, Albena, Bulgaria, 18–24 June 2015.

10. König, A.; Nicoletti, L.; Schröder, D.; Wolff, S.; Waclaw, A.; Lienkamp, M. An Overview of Parameter and Cost for Battery Electric Vehicles. *World Electr. Veh. J.* **2021**, *12*, 21. [CrossRef]
11. Hua, X.; Thomas, A.; Shultis, K. Recent Progress in Battery Electric Vehicle Noise, Vibration, and Harshness. *Sci. Prog.* **2021**, *104*, 003685042110052. [CrossRef]
12. Grycan, W. Electric Vehicles in Mining for the Aspect of Operational Safety. *Electrotech. Rev.* **2022**, *1*, 112–115. [CrossRef]
13. Gyamfi, S.; Halim, A.; Martikainen, A. Development of Strategies to Reduce Ventilation and Heating Costs in a Swedish Sublevel Caving Mine—A Unique Case of LKAB's Konsuln Mine. *Min. Metall. Explor.* **2022**, *39*, 221–238. [CrossRef]
14. Halim, A.; Lööw, J.; Johansson, J.; Gustafsson, J.; Van Wageningen, A.; Kocsis, K. Improvement of Working Conditions and Opinions of Mine Workers When Battery Electric Vehicles (BEVs) Are Used Instead of Diesel Machines—Results of Field Trial at the Kittilä Mine, Finland. *Min. Metall. Explor.* **2022**, *39*, 203–219. [CrossRef]
15. Szlązak, N.; Korzec, M.; Cheng, J. Using Battery-Powered Suspended Monorails in Underground Hard Coal Mines to Improve Working Conditions in the Roadway. *Energies* **2022**, *15*, 7527. [CrossRef]
16. Tekin, A. Assessment of Vibration Exposure of Mine Machinery Operators at Three Different Open-Pit Coal Mines. *J. S. Afr. Inst. Min. Metall.* **2022**, *122*, 235–243. [CrossRef]
17. Erdem, B.; Dogan, T.; Duran, Z. Assessment of Whole-Body Vibration Exposure of Mining Truck Drivers. *J. S. Afr. Inst. Min. Metall.* **2020**, *120*, 547–559. [CrossRef]
18. Biały, W.; Bołoz, Ł.; Sitko, J. Mechanical Processing of Hard Coal as a Source of Noise Pollution. Case Study in Poland. *Energies* **2021**, *14*, 1332. [CrossRef]
19. Alanazi, F. Electric Vehicles: Benefits, Challenges, and Potential Solutions for Widespread Adaptation. *Appl. Sci.* **2023**, *13*, 6016. [CrossRef]
20. Lyu, W.; Hu, Y.; Liu, J.; Chen, K.; Liu, P.; Deng, J.; Zhang, S. Impact of Battery Electric Vehicle Usage on Air Quality in Three Chinese First-Tier Cities. *Sci. Rep.* **2024**, *14*, 21. [CrossRef]
21. Shoman, W.; Karlsson, S.; Yeh, S. Benefits of an Electric Road System for Battery Electric Vehicles. *World Electr. Veh. J.* **2022**, *13*, 197. [CrossRef]
22. Wang, J.; Pawar, K.S.; Jayasekara, D.C.; Thoben, K.-D. Comparative Analysis of Emissions from Battery Electric Vehicles (BEV) and Internal Combustion Engine Vehicles (ICEV): A Case Study of the UK and China. In Proceedings of the 2023 IEEE International Conference on Engineering, Technology and Innovation (ICE/ITMC), Edinburgh, UK, 19 June 2023; pp. 1–9.
23. Kumar, M.; Panda, K.P.; Naayagi, R.T.; Thakur, R.; Panda, G. Comprehensive Review of Electric Vehicle Technology and Its Impacts: Detailed Investigation of Charging Infrastructure, Power Management, and Control Techniques. *Appl. Sci.* **2023**, *13*, 8919. [CrossRef]
24. Lu, M.-H.; Jen, M. Safety Design of Electric Vehicle Charging Equipment. *World Electr. Veh. J.* **2012**, *5*, 1017–1024. [CrossRef]
25. Bołoz, Ł.; Sarecki, Ł.; Ostapów, L. Meeting Demands for a Battery-Powered Future. *Glob. Min. Rev.* **2022**, *5*, 24–30.
26. *PN-EN ISO 9612:2011*; Acoustics—Determination of Occupational Noise Exposure—Engineering Method. Polish Committee for Standardization: Warsaw, Poland, 2011.
27. *PN-EN ISO 3450:2011*; Earth-Moving Machinery—Wheeled or High-Speed Rubber-Tracked Machines—Performance Requirements and Test Procedures for Brake Systems. Polish Committee for Standardization: Warsaw, Poland, 2011.
28. *PN-EN 14253:2011*; Geometrical Product Specifications (Gps)—Inspection by Measurement of Workpieces and Measuring Equipment—Part 3: Guidelines for Achieving Agreements on Measurement Uncertainty Statements. Polish Committee for Standardization: Warsaw, Poland, 2011.

MDPI AG
Grosspeteranlage 5
4052 Basel
Switzerland
Tel.: +41 61 683 77 34

Energies Editorial Office
E-mail: energies@mdpi.com
www.mdpi.com/journal/energies

www.ingramcontent.com/pod-product-compliance
Lightning Source LLC
Chambersburg PA
CBHW061630210926
51850CB00009B/2502
9783725852215